Pragmatic Engineering and Lifestyle

Pragmatic Engineering and Lifestyle: Responsible Engineering for a Sustainable Future

EDITED BY

DAVID S-K. TING
University of Windsor, Canada

And

JACQUELINE A. STAGNER
University of Windsor, Canada

United Kingdom – North America – Japan – India – Malaysia – China

Emerald Publishing Limited
Howard House, Wagon Lane, Bingley BD16 1WA, UK

First edition 2023

British Library Cataloguing in Publication Data
A catalogue record for this book is available from the British Library

ISBN: 978-1-80262-998-9 (Print)
ISBN: 978-1-80262-997-2 (Online)
ISBN: 978-1-80262-999-6 (Epub)

ISOQAR certified
Management System,
awarded to Emerald
for adherence to
Environmental
standard
ISO 14001:2004.

Certificate Number 1985
ISO 14001

INVESTOR IN PEOPLE

To everyone who practices responsible engineering and living.

Table of Contents

About the Editors

David S-K. Ting is a Professor in the Mechanical, Automotive & Materials Engineering Department at the University of Windsor, Canada. He has been teaching for more than 25 years. He has coauthored over 160 journal papers. He is the founder of the Turbulence & Energy Laboratory, and his research focuses on Flow Turbulence along with Energy Conservation and Renewable Energy.

Jacqueline A. Stagner is the Undergraduate Programs Coordinator in the Faculty of Engineering, University of Windsor, Canada. Dr. Stagner is a professional engineer with a PhD in Materials Science and Engineering and supervises research students primarily in sustainable energy and living in the Turbulence & Energy Laboratory.

About the Contributors

Radhamanohar Aepuru is an Assistant Professor in the Department of Mechanical Engineering, Facultad de Ciencias Físicas y Matemáticas, Universidad de Chile, Santiago, Chile. He also worked as an Assistant Professor in the Department of Mechanical Engineering, Facultad de Ingeniería, Universidad Tecnologica Metropolitana, Santiago, Chile. He worked as a Principal Investigator for the FONDECYT Postdoctoral Project in the Universidad de Concepcion and Universidad Tecnologica Metropolitana during the years 2018–2021 funded by ANID National Agency for Research Development, Government of Chile. Dr Aepuru served as an invited researcher (2017–2020) in the Department of Microelectronics, Technical University of Sofia, Bulgaria, for a project funded by the Bulgarian National Scientific Fund, Ministry of Education and Science, Bulgaria. He worked as a National Postdoctoral Researcher as a Principal Investigator during the years 2017 and 2018 in the Institute of Nanoscience and Technology (INST), Mohali, India, funded by the Science & Engineering Research Board (SERB), Department of Science and Technology (DST), Govt. of India.

He has published 38 refereed SCI-indexed international journals, four book chapters, one patent, and contributed to research presentations/invited talks in several international conferences. He has participated in two research projects funded by ANID National Agency for Research Development, Govt. of Chile, as a Responsible Investigator and Sponsoring Investigator. He has two projects as an Investigator and Co-Investigator funded by Universidad Technologica Metropolitana, Santiago, Chile. He currently works as a postdoctoral researcher guiding several UG students who are working in the area of Mechanical and Materials Engineering. He is an active reviewer of various international projects (Fondo para la Investigación Científica y Tecnológica-FONCyT de la República Argentina) and international peer-reviewed journals in the area of Mechanical Engineering and Materials Science and Engineering.

Kheir Al-Kodmany is a Professor of Urban Planning at the University of Illinois at Chicago (UIC), USA. He has published several books and over 100 papers on vertical urbanism, sustainability, eco-towers, placemaking, vertical farming, eco-iconic skyscrapers, transit-oriented developments, urban design, geographic information systems, and urban data visualization. Prof. Al-Kodmany also worked for the Chicago firm Skidmore, Owings & Merrill, where he designed prominent skyscrapers around the globe.

Sholihin As'ad is an Associate Professor and the Dean of the Department of Civil Engineering, Faculty of Engineering, Universitas Sebelas Maret (UNS), Indonesia. He completed his doctoral degree from the Faculty of Civil Engineering, University of Innsbruck, Austria, in 2006 and earned his master's and bachelor's degree in Civil Engineering in 1999 and 1992, respectively, from the Bandung Institute of Technology (ITB), Indonesia.

Figen Balo is a Professor of Engineering Faculty at the University of Firat, Turkey. She has published 67 papers in scientific journals, 39 book chapters, 111 research papers in conferences. As reviewer, she reviewed more than 1,653 articles for scholarly journals. Her main research area is renewable energy, renewable building-insulation materials, optimization of energy efficiency at buildings, and MCDM for renewable energy systems.

Praveen Barmavatu is working as Assistant Professor in the Department of Mechanical Engineering, Shri Jagdishprasad Jhabarmal Tibrewala University, Rajasthan, India. He is one of the youngest doctorates in India and well known for his research contribution in the field of heat exchanging systems subjected to heavy industries. He received a Master of Technology in Thermal Engineering from JNTU Hyderabad and Doctor of Philosophy from Shri Jagdishprasad Jhabarmal Tibrewala University, Rajasthan, India. He has filed his inventions and published more than six patent grants. He has participated and presented papers in many national and international conferences. He has published many book chapters and over 28 papers in scientific journals on heat transfer enhancement, compact heat exchanger, study on heat transfer characteristics by liquid jet impingement, CFD, composite preparation, plastic recycling, IC engine cylinder liners, vertical axis domestic wind mill, renewable energy systems, etc. He has supervised (supervising) many doctoral, master's, and bachelor's student theses. He is a regular reviewer for many top-class scientific journals and also has experience in funded projects.

Rupp Carriveau is the Director of the Environmental Energy Institute and the Turbulence and Energy Lab at the University of Windsor. He is a Professor in the Civil and Environmental Engineering Department at the University of Windsor, Canada. His research activities focus on energy systems futures. He has authored and coauthored multiple peer-reviewed scientific papers and presented works at many national and international conferences.

Mihir Kumar Das is Associate Professor in the School of Mechanical Sciences, Indian Institute of Technology Bhubaneswar, India. He is known for his research contribution in the field of boiling heat transfer, thermal management of electronic devices using PCM-based cooling technology, and boiling techniques. He is also into the field of thermal management of Li-ion battery. He received his MTech and PhD from the Indian Institute of Technology Roorkee, India. He has been involved in several sponsored projects by the Science and Engineering Research Board (SERB), Centre Power Research Institute (CPRI), and Bilateral collaborative projects from Department of Science and Technology (DST). He has guided many PhD, MTech, and BTech students. He has authored several scientific publications in refereed international journals, conference proceedings, and book chapters. He is also a regular reviewer of manuscripts for peer-reviewed journals of well-known publishers and reviewed external doctoral dissertations.

Sonali A. Deshmukh is working as Assistant Professor in the Department of Mechanical Engineering, S. B. Patil College of Engineering, Indapur. She is well known for her research contribution in the field of Liquid Jet Impingement for Industrial Applications. She received a Master of Technology in Heat Power Engineering from Pune University and Doctor of Philosophy from Shri Jagdishprasad Jhabarmal Tibrewala University, Rajasthan, India. She has filed her inventions and published more than three patents. She has participated and presented papers in many national and international conferences.

Mehdi Ebrahimi received his BSc in Mechanical Engineering from Ferdowsi University of Mashhad, and his MSc and PhD in Aerodynamics from Polytechnic University of Tehran, Iran. He has published 11 ISI papers and 7 book chapters, and currently, as a research engineer works at the Turbulence and Energy laboratory of the University of Windsor, Canada.

Fazıl Gökgöz received his BSc and MSc in Engineering and earned an MBA and PhD in Management with Superior Achievement Award. He served as acting head of the Privatization Project Group and associate in numerous M&A operations at the Privatization Administration of Turkey. He served as member of the board of directors and member of the auditing board at various state-owned companies. He is a full-time Professor and Head of Quantitative Methods Division in Department of Management at Ankara University Faculty of Political Sciences. He served as rectorate management coordinator and vice dean of the faculty of political sciences at Ankara University. He has supervised numerous PhD and MBA theses. He teaches Quantitative Methods at undergraduate and graduate levels and has carried out numerous international academic publications on finance, energy, and quantitative methods.

John M. Irungu is a candidate of Bachelor of Technology in Building Construction in the Department of Civil and Structural Engineering of Masinde Muliro University of Science and Technology. His focus areas are sustainable materials, sustainable architecture, and smart urban.

Edwin K. Kanda is a Civil Engineer by profession. He is a holder of a PhD in Agricultural Engineering specializing in Irrigation Engineering from University of KwaZulu-Natal (UKZN). He also holds Master of Science degrees in Water Engineering and Project Management from Moi University and Jomo Kenyatta University of Agriculture and Technology, respectively. Dr Kanda obtained a First Class Honors in a Bachelor of Technology in Civil and Structural Engineering from Masinde Muliro University of Science and Technology (MMUST). He is currently pursuing a PhD in Civil Engineering at UKZN. He is currently a Lecturer in the Department of Civil and Structural Engineering of MMUST. He has also worked as a consultant in design and supervision of civil infrastructure projects. Dr Kanda is a member of a number of professional bodies such as the

Engineer's Board of Kenya (EBK) and International Association of Hydrological Sciences.

Esmail Lakzian is an Associate Professor at Hakim Sabzevari University (*Currently, Dept. of Mechanical Engineering, Andong National University, South Korea, es.lakzian@pyunji.andong.ac.kr) specializing in renewable energy and energy conversion planning. Dr. Lakzian has published over 130 papers on renewable energy, wet steam, energy conversion, and optimization methods. All of Dr. Lakzian's works carry the mark of excellence and have been very well received in the scientific community. For instance, 21 of his papers are among the Top 10 JCR journals.

Elizabeth Lusweti is a hydrologist skilled at developing and analyzing hydrological models, analyzing maps and data sets, writing scientific articles and publications, and evaluating water quality, water use, and surface–groundwater interaction. He has over five years of experience conducting studies and consultancy services in hydrology, hydrogeology, water resources management, and conservation. Her research and writing are rooted in environmental concerns in Kenya and cross the field of gender issues, hydro-politics, and socio-hydrology. Her recent research is on the effects of oil exploration on surface water resources. Currently, she is an instructor at the Masinde Muliro University of Science and Technology (MMUST), Department of Civil and Structural Engineering. When she isn't publishing or teaching, Elizabeth loves spending time with her family and friends or going on hikes around Kenya with her son and their dog.

Fatemeh Massah holds an MSc in Mechanical Engineering from the Isfahan University of Technology, Iran. She has acquired expertise in the area of Energy Conversion after working as a Research and Development Engineer in industry. Fatemeh is presently working as a Research Assistant in the fields of solar energy harvesting, thermal energy storage, and building energy performance in the Mechanical Engineering Department at York University.

Bukke Kiran Naik has been an Assistant Professor in Mechanical Engineering Department at NIT Rourkela from 2020. He received both his PhD (2019) and MTech (2014) from IIT Guwahati and BTech (2012) from JNTU Anantapur in Mechanical Engineering. He worked as a Queen Elizabeth Postdoctoral fellow at Simon Fraser University, Canada, during 2019–2020 and as Project Engineer at IIT Guwahati in 2019. He is the recipient of Queen Elizabeth Scholars (QES) Fellowship from Universities Canada, CCSTDS Travel fellowship from INSA/CSIR/DAE-BRNS-CCSTDS, and SERB-ITS travel grant from DST, Government of India. Further, he was selected as Young Indian Scientist to the sixth BRICS (Brazil, Russia, India, China, and South Africa) conclave 2021 in the thematic area of Energy Solutions, and he also received an INAE fellowship for working as a visiting researcher at IIT Kanpur. He has filled three patents till date and secured three funded research projects from ISRO, SERB, and NIT Rourkela. He also received funding from several funding agencies (SERB, ATAL, and ISHRAE) to organize outreach programs (workshops and short-term courses). He served as a student's activities chair, K-12 chair, and CWC member for

ISHARE Guwahati subchapter from 2016 to 2018 and Program chair for ISHARE Bhubaneswar subchapter from 2021 to 2022. Currently, he is serving as ISHRAE Bhubaneswar Subchapter President for the year 2022–2023. He has about 50 research articles in reputed international journals and conference proceedings. Moreover, Dr. Naik has been invited by various reputed institutes/societies to deliver talks on his research work carried out in the energy–building–water nexus or sustainable energy and buildings.

Francis N. Ngugi is a candidate of Bachelor of Technology in Building Construction in the Department of Civil and Structural Engineering of Masinde Muliro University of Science and Technology. He is passionate about sustainable buildings and the concept of smart cities.

Paul G. O'Brien is an Associate Professor in the Department of Mechanical Engineering, York University, Canada, and has extensive experience working in leading research labs based in materials, mechanical, electrical, and chemical engineering. His interdisciplinary research efforts focus on the application of expertise from these fields to develop solutions for the realization of a transition to clean energy. Dr. O'Brien has authored over 40 papers in the areas of materials and energy. He is the founder of the Advanced Materials for Sustainable Energy Technologies Laboratory at York University, Canada, where he and his students continue to innovate and advance clean energy solutions.

Bernard O. Omondi holds a Doctorate Degree in Engineering (Applied Sciences) from Vrije Universiteit Brussels (VUB), Belgium. His areas of academic and research interest are in the fields of Affordable Housing, Green building designs, Structural Health Monitoring, Emerging Construction Materials, Advances in Concrete Technology, and Structural Design Based on Eurocodes. His major scholarly contribution was his role in developing a methodology for crack detection in structural concrete using combined Digital Image Correlation and Acoustic Emission with practical application in Prestressed Concrete Railway Sleepers. He currently serves as the Departmental Postgraduate Coordinator and Departmental Final Year Projects Coordinator. Nationally, he is a member of the technical committee on Eurocodes with Kenya Bureau of Standards. Professionally, Dr. Omondi is a registered Graduate Engineer with Engineers Board of Kenya (EBK).

Ariva Sugandi Permana earned his PhD in Urban Planning and Environmental Management from the Asian Institute of Technology, Thailand. He is currently an academic staff of the Department of Civil Engineering, School of Engineering, King Mongkut's Institute of Technology Ladkrabang. Ariva's research interests include water resources engineering, water security, water-sensitive urban design, climate change associated with water disasters, and waste management.

Chantamon Potipituk was born in April 22, 1977. She was a PhD graduate in the field of Urban Environmental Management, at the School of Environment, Resources, and Development (SERD), Asian Institute of Technology, Thailand. She is an Assistant Professor in Architecture at the Rajamangala University of

Technology Rattanakosin, Thailand. Her research interests include urban environment, sustainable development, environmental impact assessment, low carbon communities, and energy saving.

Alety Shivakrishna has been an Assistant Professor in Civil Engineering Department at Anurag Engineering College since 2014. He received an MTech (2016) from JNTUH and BTech (2013) from JNTUH in Civil Engineering. He is a research scholar in JNTU Hyderabad, Telangana. He has published eight papers on concrete partial replacements, GHG emissions, level of service, volume to capacity ratio, neuran system, etc. and presented papers in various scientific conferences.

Vineet Singh Sikarwar is a joint researcher at the Institute of Plasma Physics of the Czech Academy of Sciences in Prague, Czech Republic, the Department of Power Engineering of the University of Chemistry and Technology in Prague, Czech Republic, and the Department of Green Chemistry and Technology of the Ghent University in Belgium. He is working on the experimental and theoretical investigations of thermal plasma assisted waste valorization with the recovery of useful energy and value added materials. His other research interests include carbon capture, biofuels synthesis, and solar desalination systems. He has published his works in journals such as *Energy and Environmental Science* and *Progress in Energy and Combustion Science* among others. He is a regular reviewer for several prestigious scientific journals (e.g., *Journal of Cleaner Production, Environmental Science and Technology, Renewable and Sustainable Energy Reviews, Environment Development and Sustainability, International Journal of Chemical Reactor Engineering*). Prior to his career in research, he served as a Project Engineer in a leading heat exchanger manufacturing industry in India.

Harmeet Singh holds an MSc in Mechanical Engineering from York University, Canada. Prior to acquiring his MSc he was a full-time lecturer at Doaba Polytechnic College, India, where he was responsible for content development and teaching courses such as Machine Design, Manufacturing Technology, and Engineering Design. Harmeet presently applies his knowledge and skills working in the building sector.

Lutfu S. Sua is an Associate Professor at the Southern University and A&M College, USA. He obtained his PhD from the University of Mississippi, USA, in 2005. He also has an MBA from Troy University, USA. His primary research areas are quantitative analysis and mathematical programming, renewable energy systems, industry 4.0, and supply chain management. He has over 120 publications in international academic journals and conference proceedings within these fields.

Rathod Subash is an Associate Professor in the Malla Reddy Institute of Engineering and Technology, Medchal, Telangana, India. He is known for his research contribution in the field of thermal analysis and crack propagation of a gas turbine blade with reference to nanocoatings. He received his Master of

Engineering from University College of Engineering, Osmania University, and PhD from Shri JJT University, Rajasthan, India. He has been involved in several sponsored projects by the Science and Engineering Research Board (SERB). He has guided many MTech and BTech students. He has authored several scientific publications in refereed international journals and conference proceedings. He is also a regular reviewer of manuscripts for peer-reviewed journals of well-known publishers.

Samuel G. Waweru holds a doctorate degree in Highway Engineering from Masinde Muliro University of Science and Technology. He is an established scholar and a researcher with over 20 years' university teaching experience in Civil Engineering. His areas of academic and research interest are in the fields of highway engineering – asphalt pavements, low-cost building materials, and construction management. He has published in many international journals and participated in local and international conferences and workshops. His major scholarly contribution was his research on causes of structural failures of asphalt pavements on Kenyan highways. His current research is on the use of low-cost materials to construct affordable houses in line with the Big four agenda. He currently serves on various boards of the university and is also the current Director of Technical Vocational Education and Training (TVET) at MMUST.

Hoong Sang Wong graduated with a PhD in Applied Economics and specialized in the areas of Fisheries Economics. He had published a couple of articles in leading international journals: *Fisheries Regulation: A Review of the Literature on Input Controls, the Ecosystem, and Enforcement in the Straits of Malacca of Malaysia (Fisheries Research)* and *Bioeconomic Approach for Assessing the Status of Trawl Fishery in the Straits of Malacca (Asian Fisheries Science)*. He continues to provide expert advice in the areas of fisheries conservation to fishermen residing in the central part of the Straits of Malacca, Malaysia.

Engin Yalçın received a BA in Management from Gazi University in 2014 and an MBA from Pamukkale University in 2017. He continues to his PhD at Ankara University Department of Management and he is a research assistant at Ankara University Social Sciences Institute.

Shima Yazdani is a waste management and energy conversion planning researcher at the Hakim Sabzevari University, Iran. Dr. Yazdani has published over 17 papers on waste management, energy evaluation and sustainability, energy conversion, and optimization methods.

Chen Chen Yong is currently an Associate Professor at the Faculty of Business and Economics, University Malaya. Her research interests include human capital and applied economics. She has published several papers in international journals such as *Singapore Economic Review, International Trade and Economic Development, Chinese Business Review, Malaysian Journal of Economic Studies, Journal of Developing Areas, International Journal of Economics and Management, Fisheries Research, Asian Fisheries Science*, and so on. She has undertaken several consultancies for PricewaterhouseCoopers (PwC), PE Research, Talent Corp

Malaysia, Construction Industry Development Board (CIDB), Institute of Labour Market Information and Analysis (ILMIA), Economic Planning Unit, Malaysian Science and Information Technology Centre (MASTIC), Faradisse High Sdn. Bhd, and the Malaysian Bureau Labour Statistics.

List of Contributors

Radhamanohar Aepuru	Universidad Tecnológica Metropolitana, Chile
Kheir Al-Kodmany	University of Illinois at Chicago, USA
Sholihin As'ad	Universitas Sebelas Maret, Indonesia
Figen Balo	Firat University, Turkey
Praveen Barmavatu	Universidad Tecnológica Metropolitana, Chile
Rupp Carriveau	University of Windsor, Canada
Mihir Kumar Das	Indian Institute of Technology, India
Sonali A. Deshmukh	Shri Jagdish Prasad Jhabarmal Tibrewala University (SJJTU), India
Mehdi Ebrahimi	University of Windsor, Canada
Fazıl Gökgöz	Ankara University, Turkey
John M. Irungu	Masinde Muliro University of Science and Technology, Kenya
Edwin K. Kanda	Masinde Muliro University of Science and Technology, Kenya
Esmail Lakzian	Andong National University, South Korea
Elizabeth Lusweti	Masinde Muliro University of Science and Technology, Kenya
Fatemeh Massah	York University, Canada
Bukke Kiran Naik	National Institute of Technology, India
Francis N. Ngugi	Masinde Muliro University of Science and Technology, Kenya
Paul G. O'Brien	York University, Canada
Bernard O. Omondi	Masinde Muliro University of Science and Technology, Kenya

Ariva Sugandi Permana	King Mongkut's Institute of Technology Ladkrabang, Thailand
Chantamon Potipituk	Rajamangala University of Technology Rattanakosin, Thailand
Alety Shivakrishna	Anurag Engineering College, India
Vineet Singh Sikarwar	Institute of Plasma Physics of the Czech Academy of Sciences, Czech Republic
Harmeet Singh	Westinghouse Electric Company, Canada
Lutfu S. Sua	Southern University and A&M College, USA
Rathod Subash	Mallareddy Institute of Engineering & Technology, India
David S-K. Ting	University of Windsor, Canada
Samuel G. Waweru	Masinde Muliro University of Science and Technology, Kenya
Hoong Sang Wong	UCSI University, Malaysia
Engin Yalçın	Ankara University, Turkey
Shima Yazdani	Hakim Sabzevari University, Iran
Chen Chen Yong	Universiti Malaya, Malaysia

Book Description

Pragmatic Engineering and Lifestyle draws together international experts from engineering and architecture to disclose the latest insights into forging viable means to sustain tomorrow's needs. It focuses on breaking through barriers and fully realizing promising remedies by explicitly including the social aspect in the equation. The best way to engage the entire society is to involve them in the development and execution of the solutions. This book covers, among other topics, simple and responsible engineering and living, ecological and socially friendly buildings and infrastructures, socially resilient farming, and agroecology. This is an indispensable volume for tomorrow's engineers, architects, and policymakers. Veridically, every soul should be acquainted with, and be part of, *Pragmatic Engineering and Lifestyle.*

Preface

Tall buildings will continue to be an integral part of tomorrow's society. Pragmatically, measures must be taken to design them to be socially, economically, and environmentally sustainable. Al-Kodmany maps out the unsustainable aspects and furnishes the remedies for improving the sustainability of tall buildings whenever and wherever they are constructed in Chapter 1, "High-Rise Developments: A Critical Review of the Nature and Extent of their Sustainability."

Replacing fossil-fuel-based fibers in engineering applications such as reinforcing building materials is inevitable. The selection of the most appropriate natural fiber for a specific task with the required engineering properties such as strength, on the other hand, is not a simple process. In Chapter 2, "Application of Expert Decision Systems for Optimal Fiber Selection for Green Building Design Components," Balo and Sua present an approach for selecting the best fiber. For this green building case study, a reference is established using materials from different places.

To sustain a thermally comfortable living standard, traditional air conditioning systems must be transformed into renewable-energy-driven ones. Deshmukh et al. present the latest "Advances in Solar driven Air Conditioning Systems for Buildings" in Chapter 3. The emphasis is on reducing energy usage and, for this, solar adsorption is a promising method as it is a green technology that uses solar energy for driving the cycle.

For most developed countries, heated water is a life necessity. Tapping into solar energy can result in substantial savings in heating bills, both financially and environmentally. Singh et al. present the use of water-based Trombe walls, along with appropriate thermal energy storage, for providing heated water in Chapter 4, "Evaluating Water-based Trombe Walls as a Source of Heated Water for Building Applications." They find superior performance in terms of supplying heated water for building applications and reducing the heating load during typical Canadian winter days.

Gökgöz and Yalçin present "Investigating Waste Management Efficiencies and Dynamics of the EU Region" in Chapter 5. We cannot live responsibly if we do not manage the waste that we produce responsibly. The European Union scenario is analyzed using Slack Based Measure and Super Slack Based Measure in this chapter. Reuse and recycling resonate, and zero-waste-oriented policies are effective tools for increasing waste management efficiency.

A case study of a zero-waste-to-landfill site is disseminated in Chapter 6, "The Multipronged Approach of Solid Waste Management toward Zero Waste to Landfill Site: An Indonesia and Thailand Experience," by Permana et al. It is critical that we understand 'zero waste,' meaning no waste to the landfill rather than no waste produced. To dump no waste in one's backyard, a multipronged approach involving all stakeholders from all levels of government to every individual is required. It is found that the waste management activities realized by individual households and communities are the prerequisite for the upper levels of governing bodies.

Another way to look at zero waste is conservation, that is, start with waste reduction and proceed to zero waste. Yazdani and Lakzian detail the creation of measures to limit waste volume and toxicity in Chapter 7, "Conservation; Waste Reduction/Zero Waste." The idea is to preserve and recover resources rather than burying or burning them. A pragmatic approach is the engineering of cradle-to-cradle products, where the end of one product becomes the beginning of another. As expected, zero waste can only be accomplished with the participation of all stakeholders.

Kanda et al. enlighten us with "Adoption of Green Building Practices in Kenya: A Case of Kakamega Municipality" in Chapter 8. The aim of the study is to further green building adoption in developing countries. To do so, a better understanding of the state of green buildings is realized via a detailed survey. With that, the promotion of health and well-being, along with minimal impact on the environment via green building, can be furthered with the help of incentives including legislation and certification programs.

Furthering renewable energy is a necessary element for a sustainable future. Compressed air energy storage is one of the most promising technologies for mitigating the intermittency of renewable energy and the mismatch between energy supply and demand. In Chapter 9, "Transient Thermodynamic Modeling of Heat Recovery from a Compressed Air Energy Storage System," Ebrahimi et al. expound on the transient behavior of a compressed air energy storage system with heat recovery from the compression process and using it for heating the air during the expansion phase. Heat recovery is the key to mitigating CO_2 emission while simultaneously boosting efficiency.

Healthful nourishment is essential to sustain tomorrow and, thus, responsible fishing is included as Chapter 10, "Trawl Fisheries Management and Conservation in Malacca Straits." In this chapter, Wong and Yong convey a systematic analysis of trawl fisheries management and conservation measures in Malacca Straits. They highlight resolving of unclear and conflicting national and international territorial waters rights, strengthening national and multilateral research collaboration, employing effective and cost-saving ecological tools to identify and expand marine-protected area, and other measures to further sustain marine fisheries.

Acknowledgments

It has been a pleasure working with the wonderful Emerald publishing team; this goes all the way back to the summer of 2021 with Kimberley Chadwick. Also, a big shout-out goes to Gabriella Barnard-Edmunds, Tony Roche, Hemavathi Rajendran, Lydia Cutmore, and those working behind the scenes. We are most grateful to the 32 experts around the world who contributed the 10 chapters that made up this volume. Also invited to savor the fruits of this labor are the anonymous reviewers who ameliorated the manuscripts. Above all, providence from above sustained this endeavor from start to finish.

David S-K. Ting & Jacqueline A. Stagner
Turbulence and Energy Laboratory, University of Windsor

Chapter 1

High-Rise Developments: A Critical Review of the Nature and Extent of Their Sustainability

Kheir Al-Kodmany

Abstract

This chapter outlines complex and conflicting issues related to designing tall buildings. It gathers a vast amount of fragmented criticism and concerns and organizes them around the three pillars of sustainability: social, economic, and environmental. Mapping out the "unsustainable" aspects forms the foundation for addressing them in future research and tall building developments. The chapter engages the reader with a preliminary discussion on potential solutions to the outlined problems. It also balances extensive criticism by highlighting the virtues and advantages of tall buildings. Consequently, this chapter forms a foundation for improving the sustainability of tall buildings whenever and wherever they are constructed.

Keywords: Sky living; economic viability; social well-being; environmental safety; vertical urbanism; urban density

1. Introduction

1.1 Tall Building Construction Boom

In recent years, urbanization has been equated with building massive high-rise developments. This phenomenon is true in Asian cities (e.g., Shanghai, Beijing, Shenzhen, Tianjin, Tokyo, Mumbai, Taipei), Middle Eastern cities (Dubai, Abu Dhabi, Doha, Istanbul); European cities (London, Warsaw, Moscow), and American cities (e.g., New York, Chicago, Miami). The driving forces for building tall include massive migration from rural to urban areas, rapid urban renewal, skyrocketing land prices, active agglomeration, globalization and global competition, human aspiration, symbolism, and ego. Indeed, in dense places

Pragmatic Engineering and Lifestyle, 1–20
Copyright © 2023 Kheir Al-Kodmany
Published under exclusive licence by Emerald Publishing Limited
doi:10.1108/978-1-80262-997-220231001

where available sites for development are scarce and small, developers and landowners have no choice but to build up. Simultaneously, the increasing urban population puts tremendous demand on limited desirable places, reinforcing the need to build up. With rapid urbanization, high-rises provide various options for housing, retail, entertainment, offices, services, and amenities (fitness and health clubs, restaurants and cafes, cultural spaces, museums, etc.), allowing a city to grow without sprawling into green space and farmland. High-rise developments are becoming the global model for managing the growth of cities that need to accommodate one million incomers every week (Al-Kodmany, 2020).

1.2 Purpose of the Study

Upon scanning the social science, architecture, and planning literature, we find a plethora of scholars who critique tall buildings (Alexander et al., 1977; Ali & Moon, 2018; Barr, 2018; Kunstler, 1993; Yeang, 2022). This chapter intends to identify, collate, and consolidate fragmented concerns and critiques of tall building developments and presents them in an accessible manner. It aims to help architects and planners to attain higher levels of sustainable tall building developments by avoiding and addressing common "unsustainable" aspects. This chapter forms a knowledge base for learning and examining unsustainable practices in tall buildings. It offers a "checklist" of topics and issues important to the sustainability of tall building development. It alerts about critical and unexamined issues or provides a reminder of pitfalls and ill practices. The chapter employs sustainability as a framework to consolidate critiques and pitfalls of tall building developments and uses sustainability's three pillars (social, economic, and environmental) to guide the discussion (Du et al., 2016; Safarik et al., 2016; Yeang, 2022).

1.3 Sustainability as a Framework

Sustainability is a buzzword and a current policy, planning, and grant-writing trend. Undoubtedly, urban sustainability continues to help guide and support architecture and urban developments. The United Nations has been using sustainability to shape its universal agenda. In 2015, it adopted the 2030 Agenda for Sustainable Development, which details 17 Sustainable Development Goals (SDGs) and 169 Actionable Targets to be realized by 2030. In particular, Goal #11 refers to creating sustainable cities and communities. Further, the United Nations World Urban Forum (WUF), the world's premier conference on urban development, has embraced "sustainability" as an overarching theme for its agendas. The commitment to SDGs has been apparent since WUF's first meeting in 2002, titled "Sustainable Urbanization," in Nairobi, Kenya, through the latest in 2020, titled "Cities of Opportunities: Connecting Culture and Innovation," in Abu Dhabi, United Arab Emirates. The meeting attendance has increased from 1,200 in 2002 to 13,000 in 2020. Other important organizations, such as the World Bank, the Global Environment Facility (GEF), Local Government for

Sustainability (ICLEI), and Global Platform for Sustainable Cities (GPSC), have worked on and supported local and global sustainability projects, initiatives, and programs (Al-Kodmany, 2022).

This research uses "sustainability" as an overarching theme that guides the literature review, analysis, and synthesis. Sustainability offers a comprehensive framework represented by its three pillars (social, economic, and environmental). It seeks to balance these three dimensions. Its tripartite structure is illustrated via three intersecting circles (social, economic, and environmental) with sustainability at the intersection, emphasizing that sustainability strives to balance competing forces, needs, wishes, profits, and the availability of resources. Sustainability is a valuable framework for analyzing issues separately and collectively. Researchers have been increasingly using it to synthesize complex problems and phenomena. Some studies benefit from such a framework to compare analysis according to short- and long-term goals. Others use the framework to compare research and findings across geographic scales – from individual habitats to neighborhoods, communities, cities, regions, countries, continents, and planet Earth (Al-Kodmany, 2018; Pruetz, 2012). These pillars or dimensions are also expressed by the 3Es (equality, economics, and ecology) or the 3Ps (people, profit, and planet). In this model, "people" refers to community well-being and equity; "profit" refers to economic vitality; and "planet" refers to the environment and resource conservation. They are also known as the triple bottom line TBL or 3BL. Overall, sustainability offers a helpful universal framework for examining issues of various kinds. This research uses sustainability to discuss topics related to high-rise living (Fig. 1).

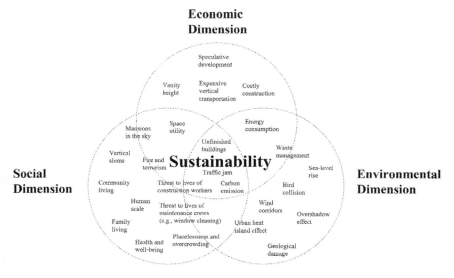

Fig. 1.   Sustainability Provides a Valuable Framework for Organizing and Discussing Issues Related to Tall Building Developments.

2. Social Dimension

Social science literature reveals that people have multiple concerns about high-rise living. Scholars highlight issues related to suitability for family living and raising children, neighborly relationships and helpfulness, personal behavior and comfort, perception of safety, tenants' relation to outdoor spaces, and connection to street life. These environments may make inhabitants feel claustrophobic, creating a rat-cage mentality. Further, high-rise living could promote poor interpersonal relationships and weak neighborly interactions, resulting in psychological depression. Further, scholars argue that low-rise living is closer to nature and facilitates a more robust community-oriented social life. As structures grow taller and taller, tenants may perceive that they become increasingly out of touch with city life (Argent, 2008; Bee & Im, 2016; Delmelle et al., 2013; Kearns et al., 2012).

2.1 Family, Community Living, and Well-Being

Sky living disconnects residents from the outside world, making them feel like they live in social isolation. Internally, they may experience social anonymity, confinement, and crowdedness. Research shows that people living in tall buildings experience more "mental health problems, higher fear of crime, fewer positive social interactions, and more difficulty raising their children" (Barr, 2018). Serious mental health problems have been correlated to building height. For example, mothers who lived in high-rises suffered more from depressive symptoms and psychological issues than those in low-rises.

Many studies found that tenants in tall buildings experience emotional stress and other adverse psychological conditions. Research indicates that undesirable social interaction among tenants because sharing floors and amenities creates pressure and tensions. In addition, the greater the degree of sharing space and utilities, the greater the stress level. Further, the high density of a building population, poor design and layout, high traffic of people in and out, and the lack of outdoor recreational and social spaces are likely to exacerbate these problems. In particular, tall buildings in poor neighborhoods suffer from a high concentration of population, overcrowding, tiny outdoor and social spaces, and a high degree of space and utility sharing. In contrast, tenants of high-end high-rises may suffer from isolation and loneliness (Gifford, 2007; Govada, 2011; Holland, 2007).

Children suffer from the feeling of living in a locked-in environment. They lack a spontaneous play environment on the upper floors that fosters exploration of the world, which is essential for their growth. Urban psychologists suggest that children (ages between two and seven) become independent by permitting them to leave to experience the natural world progressively (e.g., corner stores, shops, playgrounds, alleyways, neighborhoods, streetscapes, gardens, friends, and neighbors) and return home. This interplay between dependence and autonomy that earns a child a sense of competence is missing in a high-rise environment (Hubbard et al., 2004).

Overall, the problems of raising children in high-rises vary from essential child development matters to daily activities such as playing. One research concluded that infants' development above the fifth floor is prolonged compared to those raised below. Another study has concluded that the progress of many skills, such as clothing, assisting, and proper urination, was sluggish. As is expected, kids who reside in upper floors play outdoors less often than those in lower floors. Consequently, families with small children who can afford to live in single-family homes usually move to the suburbs.

Further, high-rises are often the habitats of smaller household sizes (the number of individuals living in a household) with fewer children critical for promoting a sense of community (McKenzie-Mohr, 2011; Worrel, 2012). High-rise residents may be afraid of heights or that a child or a loved one will jump from a window. They may fear strangers living close to their apartments or not being able to escape the building in case of fire, terrorist attack, or earthquake. Post-COVID-19, high-rise residents may fear becoming ill from infectious diseases generated by the masses living there (Delmelle et al., 2013; Gifford, 2007).

2.2 Human Scale, Placelessness, and the Public Realm

Because of their great heights and sizes, tall buildings, by default, harm human scale – they are likely to cause passersby to feel small, dwarfed, and irrelevant (Jacobs, 1963). Pedestrians at the street level are often unable to connect visually with high-rise tenants, architecture, ornamentation, decorative art, and personalized details. For example, pedestrians cannot see the flowerpots in the upper-story windows, hindering a human touch. Pedestrians cannot ultimately view the high-rise building; instead, they see "urban canyons" that make them feel visually disoriented (Gehl, 2010; Krier, 2009; Kunstler, 1993).

Similarly, tall buildings have often contributed to the problems of placelessness because of their massive size and great height. In central business districts (CBDs), tall buildings frequently evoke the image of a nerve-racking, workaholic business environment. Speaking about Manhattan, Robert Freedman (2014) contrasts high-rise with low-rise neighborhoods on the same island. He explains that in walk-up apartment neighborhoods in Manhattan, a resident or a passerby would immediately feel a warm welcome not found in the towering, elevator-skyscrapers developments that proliferate through most of Manhattan. Freedman argues that vernacular brick, wood, and stone low-rise neighborhoods are more humane than glittering, steel-and-glass high-rise neighborhoods. Tall building developments drastically alter vernacular and traditional environments (Evans et al., 2003; Kendig & Keast, 2010).

Tall building developments often hurt the public realm. Towers function as singular, autonomous structures – some are experienced as lonely sculptural objects in the cityscape. They usually do not contribute to the public sphere because they are self-contained, introverted, and privatized. Often, tall buildings require significant parking structures. Since it is costly to accommodate them underground, architects often place them above ground, thereby taking away from the street social life and unstimulating the public realm. Parking garages above ground are a "street killer" because they disconnect the social life of urban space, engender spatial disorders, and create "eyesores" in the city. Further, tall buildings adversely affect the microclimate due to wind funneling and turbulence around their bases, causing discomfort to pedestrians. These buildings cast shadows on nearby buildings, streets, parks, and open spaces, and they may obstruct views, reduce access to natural light, and prevent natural ventilation (Wood & Salib, 2013).

Importantly, current practices of tall buildings make cities worldwide look alike. For example, downtown Melbourne looks similar to Pudong, Shanghai, Miami, or Dubai. The expected shortfall of these skyscrapers is that the design has not paid attention to local tradition, geography, and climate. In particular, the steel-and-glass tower, which invaded cities, has made them look homogeneous and similar, ignoring local identity and culture.

Aesthetically, tall buildings often create contextual problems when placed near historic structures. As cities become denser and land values skyrocket, older historic structures are threatened to be demolished to make room for new taller buildings. This problem accelerates as demand for space increases and developable lots are progressively scarce. Tall buildings also could negatively affect the neighborhood character and the city skyline. Overall, historic preservation issues are often contentious and require an interdisciplinary team to make sound decisions (Ascher & Uffer, 2016).

2.3 People's Choice, Fit, and Comfort

Many governments have used high-rises to house masses of residents. For example, in the United States, from the 1960s to the 1980s, the federal government used this building typology for "affordable housing." Many residents were resentful of this living, but they had no choice. The problems in these buildings progressively became worse and worse, rendering them "vertical slums." Crime prevailed, and authorities had to take them down. Archetypal projects include the Pruitt–Igoe in Saint Louis, Missouri, and Cabrini Green in Chicago, Illinois. In other incidences, residents would refuse to live in high-rises in the first place. For example, the Chinese government has recently constructed high-rise cities to house villagers. However, they shunned these places, favoring low-rise living. Chinese people have nicknamed these cities "modern ghost cities." Ordos Kangbashi in China is one of several new high-rise cities that suffered from this problem (Al-Kodmany, 2018).

Using high-rise buildings as affordable housing has often featured poor design and living conditions. As we see in Section 3, tall buildings are costly by default. The only way to reduce costs is to compromise on the quality of building materials, finishes, interior design, appliances, amenities, services, and the like. The US authorities demolished public housing high-rises partially because of poor design. Overall, tall buildings require high maintenance, and any malfunctioning has a rippling effect. For example, a water leak will affect the floor where the problem occurs and all the lower floors. Likewise, fire has an upward rippling effect as it spreads vertically rapidly. Wall cracks are likely to expand and threaten larger portions of the building. Unexpected foundation settlement may endanger the entire structure. Elevator malfunction is also a common problem, which has a rippling effect on the ground floor lobby, where people gather most, and throughout the building. Usually, people exhibit anxiety and discomfort after waiting 30 seconds in a commercial building and 45 seconds in a residential building.

It is important to note that ultraluxury and expensive tall buildings are not immune to these problems. For example, 432 Park Avenue of New York City in the so-called Billionaire's Row, the residence of some of the world's wealthiest and most potent, recently faced "water damage from plumbing and mechanical issues; frequent elevator malfunctions; and walls that creak like the galley of a ship,... Residents at 432 Park complained of creaking, banging, and clicking noises in their apartments and a trash chute 'that sounds like a bomb' when garbage is tossed" (Chen, 2021).

2.4 Construction, Repair, and Maintenance

Constructing tall buildings, particularly supertalls, could be risky. Over the years, thousands of workers lost their lives because of accidents, equipment malfunction, or hazardous working conditions. Unfortunately, building tall buildings relies on workers who manually carry out duties. For example, façade assembly and exterior cladding depend on workers who grab panels from cranes and place them in assigned spaces. Workers repeat this process until they complete each façade. These tasks are more hazardous on upper floors where wind velocity is high. As architects design more complex forms and shapes of buildings (including tall ones), construction workers face more significant risks (Bao et al., 2015).

Similarly, routine repair and upkeep of tall buildings risk workers' lives. For example, tall buildings' window cleaning remains a frequent reason for the laborers' demise, who carry out duties by gradually falling the building's total height while dangling on ropes and holding water containers and washing tools. Some workers daze and fall, while others bounce into walls and windows due to forceful wind. Overall, the task could be daunting. For example, it takes 36 workers four months to wash Burj Khalifa's 26,000 windows (Al-Kodmany, 2015).

Further, window cracking and breaking are common problems in supertall buildings. Indeed, glass ages and weakens over time, and any deficiencies in manufacturing or installation could lead to cracks or breakups under wind pressure. For example, Willis Tower (formerly Sears Tower) in Chicago has experienced several incidences where some windows on the upper floors were shattered under a forceful wind. Debris fell on sidewalks, damaged properties, and hurt pedestrians. For safety considerations, police blocked streets and rerouted traffic, causing inconveniences and traffic congestion in adjacent neighborhoods. Reinstalling glass and repairing cracks are also formidable tasks. Further, elevators require close monitoring and maintenance. As such, building managers need to hire resident engineers who should be experienced in mechanical and electrical systems, IT networks, software, and programming languages.

3. Economic Dimension

Skyscrapers are costly buildings. Their costs are higher than that of low-rise buildings holding the same square footage due to their high embodied energy, i.e., the resources needed to build them. They need more robust structural systems to withstand the natural forces of wind, gravity, and earthquakes and resist severe weather conditions such as hurricanes, tornados, and typhoons. Tall buildings demand enormous amounts of steel and concrete and require expensive vertical transportation such as elevators and escalators and tremendous energy to pump water to upper floors. They suffer from diseconomies of vertical construction systems (e.g., taller cranes, jumping cranes, "kangaroo cranes," jumping boards, and hydraulic pistols). Pumping concrete to higher floors demands powerful pumps and special concrete that can travel long distances without stiffening too soon, resulting in clogging hoses. Skyscrapers also feature a lower "net-to-gross" ratio referring to the net useable space in the building – about 70% for high-rise buildings compared to more than 80% for low-rise buildings (Al-Kodmany & Ali, 2016; Ali & Moon, 2018; Du et al., 2016).

3.1 Space Efficiency

High-rise buildings use space less efficiently than low-rise buildings. For example, they need many elevators simply because they are the prime mode of trans-portation – people are usually unwilling to walk up more than a few floors. Second, people do not tolerate long waits. Therefore, engineers compute the needed number of elevators so that tenants do not wait for elevators for more than a certain number of seconds – about 30 seconds for commercial office buildings and 45 seconds for residential ones (Eappen, 2017). Indeed, if an elevator malfunctions, overcrowding develops quickly in the lobby. Skyscrapers also need multiple elevator types (e.g., local, express, service, freight, firefighters). Therefore, elevators add high costs to the building and consume significant useable space. Post 9/11, authorities have placed more stringent requirements on all vertical transportation. For example, new codes require stairways to be wider to accommodate two flows of people – a flow of tenants escaping the building and a flow of firefighters going up to rescue tenants. Overall, as we are building higher, architects must incorporate advanced elevator systems to allow tenants to reach their destinations swiftly while ensuring their comfort and safety (Al-Kodmany, 2020).

   In addition to "wasting" spaces to house elevators and mechanical, structural, and damping systems, skyscrapers may "waste" areas for merely boosting height. The Council on Tall Buildings and Urban Habitat (CTBUH) has coined the "vanity height" term to refer to the wasted space between a skyscraper's highest occupiable floor and its architectural top. As such, the "vanity ratio" equals "vanity height" divided by the architectural height of the building. Skyscrapers increasingly feature a larger vanity ratio. The average vanity ratio in the UAE is 19%, making it the nation with the "vainest" tall buildings. The Burj Khalifa's vanity height is 244 m (800 ft), which qualifies as a skyscraper. Burj Al Arab in Dubai, UAE, has a 39% ratio (124 m: 321 m), (407 ft: 1053 ft) – the most remarkable "vanity ratio" among completed supertalls (Al-Kodmany, 2018).

3.2 Speculative Investment

Financially, tall building developments could be a risky investment where developers bet on economic growth and overlook economic recession that results in massive vacancies in these buildings. Often, the market experiences a housing bubble when extortionate profit potential triggers a construction boom that exceeds demand and affordability, and when the economy slows down, the housing bubble bursts. A similar situation happens in the office space sector. Office buildings suffer from vacancies simply because of the cost of running the building or because of an outdated look and functionality. Further, the fluctuating nature of financial and lending systems may delay, discontinue, or cancel the construction of tall buildings. For example, because of the 2008 financial crisis, the developer of the Spire Tower in Chicago canceled the project after completing the building's foundation. Consequently, the site remains deserted, creating an eyesore in the community.

Further, demographic changes and lifestyle shifts could challenge tall buildings' sustainability promise. For example, recent ultraluxury tall building developments in the United States have been betting on the exceedingly wealthy people who form a small proportion of the world population. Owners of these buildings sell housing units for tens of millions of dollars. However, developers are bearing the risk of overshooting the mark. Other residential tall building developments have been betting on the millennials and downsizing retirees. Nevertheless, these developments may face high vacancies when the millennials flock to suburbs to start families and the retirees' population declines (Heinonen et al., 2011).

3.3 Building Construction and Uncompleted Buildings

Further, design and construction mistakes could have a rippling effect that would prolong the construction period, thereby incurring additional costs. For example, the John Hancock Center in Chicago faced construction problems. After building the foundation, which consisted of 57 caissons (8-foot-thick (2.4-m-thick) concrete columns), workers discovered that one of the caissons had shifted 0.9 in (2.3 cm). Specialized workers had to perform sonic tests to detect weak spots. They found that the contractors removed machinery while the concrete was still settling to save time and money. Chicago's fragile soil had seeped into concrete, causing the shift. Correcting this problem set the project back several months and increased construction costs (Leventis et al., 2015).

While under construction, the 10,344 windows of the 60-story 200 Clarendon (alternative names include John Hancock Tower and Hancock Place) in Boston, Massachusetts, were cracked and replaced with temporary sheets of plywood and later with permanent thick glass – a process that delayed opening the building for 4–5 years. The original façade consisted of double-layered reflective glass with a thin strip of lead sandwiched between the two layers. The lead layer had begun to develop fatigue and crack. Further, because the lead layer glues so tightly to the glass, it transferred cracks into the reflective chrome coating on the glass, eventually causing the glass to crack (Bilaine, 2015).

Severe weather conditions and natural hazard events also impact and delay the construction process of tall buildings, adding substantial costs. For example, on October 29, 2012, when One World Trade Center was under construction, Hurricane Sandy landed on the East Coast and hit New York City, including the Ground Zero Site (One World Trade Center). Rainwater has soaked the unfinished structure and filled the 16-acre site with 10–140 feet of water depths, totaling about 125 million gallons. Construction workers had to drain water from the entire building and the site. Similarly, when Shanghai Tower was under construction, a typhoon hit the tower, causing damage and delay (Bilaine, 2015).

Noticeably, there were incidences where the construction of tall buildings started but was not completed due to financial hurdles, political pressure, and cultural opposition. Indeed, financing a skyscraper is not a simple task. Many ambitious tall projects start during economic booms, and when economic busts occur, these buildings suffer (Al-Kodmany, 2018).

4. Environmental Dimension

Skyscrapers suffer from a large carbon footprint observed in construction, operation, maintenance, and demolition at the end of their life cycles. They exert significant demand on infrastructure and transportation systems, creating overcrowding and traffic congestion. In some projects, for example, the Brickell City Center in Miami, Florida, the developer had to relocate existing infrastructure (e.g., water, sewer, and drainage utilities) to accommodate buildings' footprints, deep foundations, basements, and parking garages.

4.1 Carbon Emission

Skyscrapers' construction and operation require tremendous energy and generate significant amounts of carbon emissions and air pollution that contribute to global warming. High-rises consume lots of steel and cement – manufacturing these materials requires lots of energy and generates large amounts of carbon dioxide. Also, tall buildings construction requires massive power. It causes considerable carbon dioxide by operating heavy machinery and equipment such as powerful cranes and pumps (e.g., pumping water and concrete to upper floors) and dump trucks. Transporting building materials from far distances (sometimes across the globe) also consumes energy and produces immense carbon dioxide (Wilson et al., 2013).

Further, skyscrapers consume great energy and generate significant greenhouse emissions from running mega electrical, mechanical, lighting, and security systems. Architects have built towers with poor thermal performance and without natural ventilation, meaning that buildings' owners need to continuously heat and cool indoor spaces (in the winter and summer, respectively) to ensure tenants have a comfortable indoor environment. As such, the energy needed to heat and cool these skyscrapers is costly and hurts the environment by generating massive carbon dioxide (Elliott, 2010).

4.2 Urban Heat Island Effect

When extreme heat occurs, high-rise cities have more trouble cooling off than other places, creating a greater demand for energy to cool spaces. A concentration of tall buildings increases the city's thermal mass, and consequently, it increases the Urban Heat Island (UHI) effect. As cities grow denser and accommodate taller buildings and more significant auto traffic, UHI intensity and carbon emission will increase significantly (Giridharan & Ganesan, 2004; Sarrat et al., 2006; Wagner & Viswanathan, 2016).

4.3 Wind and Natural Ventilation

Urbanization weakens natural ventilation because buildings block breezes from nearby natural fields such as the ocean, sea, lakes, forests, farms, and mountains (Kawamoto, 2016). Given their greater heights and larger masses, tall buildings impact natural wind directions and patterns by increasing the distance of wind shadow and minimizing the airflow in the leeward directions, i.e., behind buildings. Therefore, in polluted urban environments, decreased airflow augments stagnation and accumulation of air pollution. Tall buildings create a wind tunnel effect that increases wind speed and turbulence and discomforts pedestrians at the street level. Strong airflow around tall buildings creates eddies, dust loops, and air pollution, disturbing and discomforting street activities. Wind acceleration manifests in open areas, including plazas, passages, entrances, corners, and spaces between buildings (Lynch, 2015).

4.4 Geological Considerations

The geological structure of a place poses several implications for constructing tall. For example, when tall buildings stand on thin soil, their collective weight may make them gradually sink. Shanghai offers an illustrative example. While it is safest to anchor a skyscraper's foundation over a bedrock (a geological layer of solid rock), it is not always easily accessible. In some cities, the bedrock is so deep that it would be too expensive to reach. Other cities have a combination of swampy soil and deep bedrock. Steadying skyscrapers in these places is costly because they require a staggering feat of structural engineering, otherwise, towers could collapse (Xia et al., 2010).

4.5 Bird Collision

Bird-glass collisions are an unfortunate side effect of tall building developments. Worldwide, billions of birds perish from collisions with glass yearly, making it the second-largest human-made hazard to birds after habitat loss (Petty, 2014). Since most modern tall buildings are clad in glass, tall buildings have become a prime killer. Transparent and reflective glass kills birds because birds perceive clear glass as an unobstructed passageway and, consequently, they attempt to fly through. On the other hand, reflective glass reflects the sky, clouds, and nearby vegetation, reproducing a perceived habitat familiar and attractive to birds. Birds usually use stars and the moon, and illuminated windows often divert them from their original flight paths. At night, skyscrapers' lights lure birds in search of navigational cues. As such, birds can be attracted to artificially lit tall buildings, resulting in collisions.

4.6 Waste Management

Tall buildings generate large volumes of waste because they house a large population. On average, the disposal rate of an apartment unit is about one ton per year. While this amount of waste is not different from a low-rise residential unit, the method of waste collection in high-rises is more complicated than that in low-rises. One popular disposal method for tall buildings is the chute system, which consists of vertical shafts that transfer waste to a central location bin on a lower building level via gravity. Nevertheless, the large amount of garbage accumulated on the ground floor poses a challenge to management systems. That is, massive trash can pile up quickly. If the building management does not take care of the disposal rooms, a large amount of daily waste can lead to overflowing collection bins that can cause odor and rodent problems. As such, collecting waste by haulers must occur more than once a week, which is the norm with single-family residential trash pick-ups.

While cities increasingly enact recycling goals, high-rise apartments remain mainly immune to recycling. The complex collection methods and high-rise confined spaces make it harder to implement recycling systems. Also, loading docks have limited container space, and service-parking areas are small, making it harder for trucks to maneuver, load waste, and leave the site. Overall, because of rapid urbanization, the issue of waste management is increasingly significant. Cities often lack waste-recycling facilities, regulations on the disposal of construction waste, and the management of dumping sites. Furthermore, numerous cities face illegal dumping problems and a lack of landfill sites (Parker & Wood, 2013; Safarik et al., 2016).

5. Discussions

5.1 Searching for Remedies to Problems of High-Rise Developments

The growing proliferation of tall buildings in our cities threatens their sustainability. Indeed, the previous review suggests that high-rise developments suffer from myriad challenges and issues. Since they house huge populations in mega-developments, their social, economic, and environmental impacts are immense. Therefore, architects, engineers, planners, and public officials must seek ways to better integrate these "urban giants" in their specific contexts while respecting local cultures, habits, practices, and climates.

This chapter suggests sustainability as a practical conceptual framework to examine many issues related to the ills and downsides of tall building developments. Sustainability helps to separate topics for deep examinations. Organized around three major social, economic, and environmental spheres, it covers the most critical issues related to human health, comfort, and environmental well-being. As such, the sustainability thesis applies to various problems related to high-rise development, be it planning, design, construction, social life and programming, indoor and outdoor activities, costs, operation, maintenance, and environmental impact.

High-rises cram large populations into smaller indoor places. Consequently, architects and planners should compensate for that by providing well-designed, attractive communal spaces. Small indoor spaces prohibit tenants from holding large social gatherings; hence, public spaces (indoor and outdoor) are crucial for the social well-being of occupants. Outdoor areas may offer additional benefits in bringing tenants and visitors of multiple high-rises together and supporting large-scale social and cultural events. For example, outdoor spaces may support a block party, where many community members congregate and celebrate. Carefully programming social indoor and outdoor spaces is vital to their success.

Unfortunately, high-rises fall short in space efficiencies because they demand extra spaces for vertical transportation (elevators, stairwells, escalators, fire exits), waste management ducts, wider columns, and utility floors. On average, these spaces take up about 30% of the building. Therefore, efficient use of space is a must in designing high-rises. The spatial layouts of the whole building and individual units must minimize wasted spaces. The circulation scheme should shorten travel distances. Odd forms that entail unusable spaces, dead spaces, and useless corners should be avoided.

Economically, high-rises are costly. Their construction, operation, and maintenance are more expensive than low-rise buildings. Reducing costs by attaining lower-quality buildings should be avoided. Such remedies may create poor living and working environments with negative consequences (Wells, 2014). They also face issues of high vacancy due to market changes. As such, developers, architects, and building owners may consider multischeme, mixed-use developments that ensure some building parts are continuously rented and occupied.

Renewable energy features, solar shading, and double-skin facades may reduce operational energy costs and environmental effects. However, greenwashing should not go unchecked. For example, architects and engineers have made many claims about using wind turbines as a source of renewable energy. A close examination of the issue reveals that only a few high-rises worldwide have employed wind turbines because of mainly little return on investment (ROI). Their ROIs have been low, especially in places crowded with tall buildings that block the wind. Also, buildings with integrated wind turbines face technical challenges, maintenance issues, and noise problems. A similar situation is found in claims about photovoltaic panels.

Urban design studies should be carried out to reduce the shadow effect, improve ventilation, and support biodiversity and ecology. A comfortable microclimate is needed to support social activities and the well-being of plants. The pedestrian realm should be carefully designed to match the local environment, cultural habits, and changing seasons. For example, the availability of direct sunlight to the pedestrian realm is essential when the temperature drops. Conversely, on hot days, protecting pedestrians from the sun is needed. Partial solutions include integrating into the pedestrian realm physical features, such as roof-like covers, overhangs, and awnings, to keep the sun or rain off pedestrians. These features could be attached and detached from buildings to ensure proper pedestrian protection. Architects and landscape architects should collaborate to ensure that their visual impact accentuates the original design and esthetics.

Natural ventilation solutions should be explored to reduce environmental impact and improve health conditions. This could be most applicable in mild climate areas and buildings that are not too high. For example, buildings with 10–30 stories in pleasant climates may employ natural ventilation solutions. Higher floors may not enjoy natural ventilation dues to high-speed winds. Buildings that support natural ventilation have the potential to reduce carbon emissions by 60% compared to air-conditioned buildings of similar sizes. Similarly, natural light should be harnessed to reduce environmental impact and costs. Building orientation should be considered with the sun path year-round. Further, architects should also observe a short lease span (the distance between the exterior wall and core) to increase natural light in interior spaces.

Fire is a significant concern in high-rises due to difficulties rescuing tenants. A few measures are needed to mitigate this problem. Notably, the swiftness of fire detection is critical as it allows building managers to intervene sooner, preventing further escalation. As such, employing smart detectors could help sense the presence of the tiniest smoke. Positive air pressure and adjustable ventilation systems can prevent smoke from spreading. Live messaging systems could alert tenants for safe evacuation.

Technological advances yield a broad spectrum of sensors, including lighting sensors, heating, and cooling sensors, weather sensors, motion sensors, escalator sensors, elevator sway sensors, in-cab weight sensors, ventilation and humidity sensors, CO_2 sensors, blinds, and shades sensors, parking sensors, and earthquake sensors. Overall, installing detectors can help reduce energy use and promote efficiency and safety. Sensors can also help to monitor water use and wastewater discharges. Irrigation sensors are also efficient. We may consider adding aerators and occupancy sensors on lavatory faucets to reduce water flow by mixing water with air while maintaining adequate pressure. Similarly, master switches to control lighting and appliances can save energy. A single master switch at the entrance of each unit allows residents to turn off lights in all rooms simultaneously as they leave the residence. Further, a simple replacement of regular lights with LED systems could lead to significant savings.

Cookie-cutter, poor-quality high-rise buildings should be avoided. Public housing has been stigmatized partially due to mediocre design and poor image-ability. In the United States, these projects failed because of mainly poor social life. However, the cookie-cutter image and poor design have contributed to their failure. Introducing various forms, colors, shapes, and sizes may alter that stigma about public housing. The spatial composition in the site plan could change the perceptual characteristics of development. Also, varying the heights of buildings generates a better profile and skyline. Recent public housing projects in Singapore offer good examples.

Killing birds has been a severe problem caused by tall buildings. As mentioned, hundreds of millions of migratory birds die after colliding with buildings yearly. Architects may integrate safety elements (e.g., verandas, screens, grilles, awnings, shutters, opaque glass) to mitigate the problem by deterring them from the building. Also, fritted glass with patterns of printed dots, liens, strips, marks, and labels could help deter birds. Surface reflectivity is the prime culprit. The key to all exterior materials is to reduce reflectivity so that birds do not see the building's skin as an extension of the surrounding natural environment. If they do, they will fly into the building. Three-dimensional geometric glass may play the same trick. Other tips include dimming artificial light and shutting lights at night (when not in use), drawing curtains and blinds, using desk lights instead of ceiling lights, and integrating light sensors. New York City recently passed a bird-friendly bill requiring 23-meter/75-feet buildings and above (existing and planned) to adhere to similar guidelines to reduce bird-building collisions.

Finally, with the advent of COVID, the sustainability thesis will be reinforced. COVID-19 has reminded us of the importance of natural ventilation that helps reduce the spread of the virus. Building owners and developers started to value advanced ventilation systems within tall buildings more than ever before. Examples include negative air pressure (which contains pathogens from spreading), displacement ventilation (which supplies cooler or warmer air from below and lifts contaminants), and clean air ventilation (which invites fresh air in rather than circulating inside existing stale air). In the post-pandemic era, it will be easier to make a case for water filtering systems to fight situations where a virus can contaminate the water supply. Likewise, it will be easier to make the case to the developers and owners to invest in intelligent/innovative systems that ensure high-quality air and water supply.

5.2 Tapping the Potential of High-Rise Developments

In parallel with finding solutions to problems presented by tall building developments, architects, engineers, and planners should capitalize on the advantages this building typology may offer. Indeed, tall buildings allow the clustering of a large population in a smaller land, allowing the integration of open and green spaces on the ground floor. Urban parks function as the neighborhood's lungs and offer vital social benefits. In contrast, dense low-rise developments often eat up much of the land on the ground floor, leaving little space for public parks. Usually, the large concentration of people near public parks ensures frequent visits by tenants. Likewise, clustering masses of people in smaller areas offer more significant opportunities for success for all kinds of amenities, retail, cultural facilities, entertainment, restaurants, cafes, health clubs, and the like. Concentrating many people in a smaller area guarantees significant clientele that helps sustain these services. Simultaneously, the vertical spatial arrangement saves people traveling long distances on foot or by vehicle.

In commercial urban cores and financial districts, spatial verticality supports agglomeration. Urban agglomeration and economic synergy require the proximity of businesses, and high-rise developments do just that. The large clusters of companies engender "knowledge spillovers" among agencies in the same sector and across sectors that foster innovation and creativity. In compact and dense environments, knowledge can spread into unintended disciplines, and a vital share of knowledge transfer happens spontaneously. In addition, clustering firms producing similar products spurs competition, which pushes firms to embrace efficiency, reduce prices, and improve yields. Further, agglomeration enhances the economy of scale. Buchanan's research attempted to quantify the benefits of agglomeration. His research shows that "a doubling of employment density within a given area can lead to a 12.5% additional increase in output per worker in that area. For the service sector, the figure is far higher at 22%" (Buchanan, 2008). Buchanan's research has estimated that moving 80,000 jobs in London to high-density locations could increase workers' output by £206 million (Buchanan, 2008).

Importantly, clusters of tall buildings should center around major mass transit stations. Mass transit's advantages include cutting the city's carbon footprint and reducing gasoline consumption and traffic congestion. The vertical arrangement of tall buildings allows efficient movement of transit people by shortening origin–destination distances. Vertical transportation cuts walking distances, a crucial element needed to support catchment areas and sustain ridership. Similarly, clustering tall buildings around mass transit nodes can strike down sprawl. For example, planning authorities in the Washington, DC, region have applied this concept to cut urban sprawl, exemplified by the Rosslyn-Ballston (R-B) Corridor. The 3.3-mile R-B Corridor integrates five clusters of tall buildings around mass transit nodes in five neighborhoods: Ballston, Rosslyn, Courthouse, Clarendon, and Virginia Square. Planning authorities in Washington, DC, also applied this concept to reduce suburban sprawl. They have created transit villages along metro lines. Examples include Silver Spring, Tysons Corner, North Bethesda, Bethesda, Alexandria, Reston, and Rockville. Therefore, clustering many people near mass transit nodes offers a significant advantage that planners must harness.

Tall buildings have a reputation for being energy guzzlers. This assumption could be valid as high-rises consume significant energy to heat and cool large areas. However, empirical studies have compared tall buildings' energy consumption to that of suburban developments, and results indicated comparable consumption per capita (Du et al., 2016). Demographic changes have resulted in having a similar number of household members that live in large suburban single-family homes and high-rise apartments. Middle-class and wealthy suburban homes vary from 3,000 to 10,000 square feet, while high-rise apartments feature much smaller sizes. As such, there are cases where apartments are more energy-efficient. High-rise buildings suffer from high embodied energy (energy required to build them). Still, suburban developments of large single-family homes take up lots of embodied energy by engaging and altering a larger swath of land (Al-Kodmany & Ali, 2013).

6. Conclusions

This chapter collates significant critiques of tall buildings. The reader may conclude that tall buildings are the wrong choice for urban development. In other words, the reader may misconceive this chapter as support for views that render tall buildings' developments undesirable because they are costly, wasteful, and unfriendly to the environment and social living. However, pressing factors preclude avoiding them; examples of these factors include expensive land, small developable site, strategic location, zoning regulations, demographic change, global demand, scenic views, etc. As a building typology, tall buildings are likely to stay, not go away (Yeang, 2022). Therefore, the job of architects and urban planners is to minimize the adverse effects and tap their potential, as outlined in this chapter. Thus, this chapter raises awareness of the common problems of tall buildings to search for remedial measures. It systematically organizes the issues using the sustainability framework to alert developers, architects, and urban planners about significant pitfalls. By applying the sustainability framework, we can enhance future developments of tall buildings.

7. Future Research

Overall, scholars may use this chapter as a stepping stone to future research. Future studies should examine the outlined issues in this chapter in greater depth and breadth. They should attempt to find remedies and solutions. Studies that examine social sustainability are lacking. Learning from tenants' experiences and case studies will enrich this line of research. We need to collect additional data for the sustainable performance of existing tall buildings. In particular, we continue to lack access to data on operational costs, which is crucial to know about the functioning of these buildings. Similarly, many claims about environmentally "sustainable" high-rises remain unchecked because of insufficient data.

References

Al-Kodmany, K. (2018). *The vertical city: A sustainable development model*. WIT Press.

Al-Kodmany, K. (2015). *Eco-towers: Sustainable cities in the sky*. WIT Press.

Al-Kodmany, K. (2020). *Tall buildings and the city: Improving the understanding of placemaking, imageability, and tourism*. Springer Publisher.

Al-Kodmany, K. (2022). Sustainable high-rise buildings: Toward resilient built environment. *Frontiers in Sustainable Cities, 4*, 1–13. https://doi.org/10.3389/frsc.2022.782007

Al-Kodmany, K., & Ali, M. M. (2013). *The future of the city: Tall buildings and urban design*. WIT Press.

Al-Kodmany, K., & Ali, M. M. (2016, December). An overview of structural developments and aesthetics of tall buildings using exterior bracing and diagrid systems. *International Journal of High-Rise Buildings, 5*(4), 271–291.

Alexander, C., Ishikawa, S., & Silverstein, M. (1977). *A pattern language: Towns, buildings, construction*. Oxford University Press.

Ali, M. M., & Moon, K. S. (2018). Advances in structural systems for tall buildings: Emerging developments for contemporary urban giants. *Buildings, 8*, 104. https://doi.org/10.3390/buildings8080104

Argent, N. (2008). Perceived density, social interaction, and morale in New South Wales rural communities. *Journal of Rural Studies, 24*, 245–261. https://doi.org/10.1016/j.jrurstud.2007.10.003

Ascher, K., & Uffer, S. (2016). *Tall versus old? The role of historic preservation in the context of rapid urban growth* (pp. 107–114). Council on Tall Buildings and Urban Habitat.

Bao, L., Chen, J., Qian, P., Huang, Y., Tong, J., & Wang, D. (2015). The new structural design process of supertall buildings in China. *International Journal of High-Rise Buildings, 4*, 219–226.

Barr, J. (2018, January 31). The high life? On the psychological impacts of highrise living. In *Building the skyline: The birth and growth of Manhattan's skyscrapers*. https://buildingtheskyline.org/highrise-living/#:~:text=The%20Findings&text=Across%20these%20different%20categories%2C%20a,difficulty%20with%20raising%20their%20children. Accessed on May 15, 2022.

Bee, A. S., & Im, L. P. (2016). The provision of vertical social pockets for better social interaction in high-rise living. *Planning Malaysia, 14*. https://doi.org/10.21837/pmjournal.v14.i4.156

Bilaine, C. (2015). *Challenges of building Brickell city center: A true mixed-use mega project downtown Miami* (pp. 93–100). CTBUH Research Paper; Council on Tall Buildings and Urban Habitat.

Buchanan, C. (2008, September). The economic impact of high density development and tall buildings in central business districts, a report for the BPF (pp. 1–43). *British Property Federation*. http://www.ctbuh.org/Portals/0/People/Working Groups/Legal/LegalWG_BPF_Report.pdf. Accessed on July 2022.

Chen, S. (2021, February 3). The downside to life in a supertall tower: Leaks, creaks, breaks. *The New York Times*. https://www.nytimes.com/2021/02/03/realestate/luxury-high-rise-432-park.html. Accessed on May 15, 2022.

Delmelle, E. C., Haslauer, E., & Prinz, T. (2013). Social satisfaction, commuting and neighborhoods. *Journal of Transport Geography*, *30*, 110–116. https://doi.org/10. 1016/j.jtrangeo.2013.03.006

Du, P., Wood, A., & Stephens, B. (2016). Empirical operational energy analysis of downtown high-rise vs. Suburban low-rise lifestyles: A Chicago case study. *Energies*, *9*, 445. https://doi.org/10.3390/en9060445

Eappen, R. (2017). Elevator maintenance considerations for supertall buildings. *CTBUH Journal*, 28–33.

Elliott, D. (2010, December). A useful tool with room for improvement (pp. 38–43). *Planning, the Magazine of the American Planning Association.*

Evans, G. W., Wells, N. M., & Housing, M. A. (2003). Mental health: A review of the evidence and a methodological and conceptual critique. *Journal of Social Issues*, *59*, 475–500. https://doi.org/10.1111/1540-4560.00074

Freedman, R. (2014, March 12). Mid-rise: Density at a human scale *Planetizen*. https://www.planetizen.com/node/67761. Accessed on July 15, 2017.

Gehl, J. (2010). *Cities for people.* Island Press.

Gifford, R. (2007). The consequences of living in high-rise buildings. *Architectural Science Review*, *50*, 2–17. https://doi.org/10.3763/asre.2007.5002

Giridharan, R., & Ganesan, S. (2004). Daytime urban heat island effect in high-rise and high-density residential developments in Hong Kong. *Energy and Buildings*, *36*, 525–534.

Govada, S. (2011). ULI ten principles to help guide large scale integrated development. *Urban Land Magazine.* https://urbanland.uli.org/industry-sectors/ infrastructure-transit/uli-ten-principles-to-help-guide-large-scale-integrated-development/. Accessed on May 10, 2022.

Heinonen, J., Kyrö, R., & Junnila, S. (2011). Dense downtown living more carbon intense due to higher consumption: A case study of Helsinki. *Environmental Research Letters*, *6*, 1–9.

Holland, C. (2007). *Social interactions in urban public places.* Policy Press.

Hubbard, P., Lilley, K., & Hubbard, P. (2004). Pacemaking the modern city: The urban politics of speed and slowness. *Environment and Planning D: Society and Space*, *22*, 273–294. https://doi.org/10.1068/d338t

Jacobs, J. (1963). *The death and life of great American cities.* Random House Publishing.

Kawamoto, Y. (2016). Effect of urbanization on the urban heat island in Fukuoka-Kitakyushu metropolitan area, Japan. *Procedia Engineering*, *169*, 224–231.

Kearns, A., Whitley, E., Mason, P., & Bond, L. (2012). 'Living the high life'? Residential, social and psychosocial outcomes for high-rise occupants in a deprived context. *Housing Studies*, *27*, 97–126. https://doi.org/10.1080/02673037.2012. 632080

Kendig, L. H., & Keast, B. C. (2010). *Community character: Principles for design and planning.* Island Press.

Krier, L. (2009). *The architecture of community.* Island Press.

Kunstler, J. H. (1993). *The geography of nowhere: The rise and decline of America's man-made landscape.* Simon & Schuster.

Leventis, G., Poeppel, A., & Syngros, K. (2015). From supertall to megatall: Analysis and design of the Kingdom Tower piled raft. In *Proceedings of the CTBUH 2015 International Conference,* New York, NY, USA, 26–30 October (pp. 44–53).

Lynch, E. (2015, February). Hurricane Sandy's wake-up call: The New York area redefines recovery. *Planning, the Magazine of the American Planning Association,* 25–31.

McKenzie-Mohr, D. (2011). *Fostering sustainable behavior: An introduction to community-based social marketing.* New Society Publishers.

Parker, D., & Wood, A. (2013). *The tall buildings reference book.* Routledge, Taylor & Francis Group.

Petty, T. (2014, February 12). Hundreds of millions of birds killed annually from building collisions. *Audubon.* http://www.audubon.org/news/hundreds-millions-birds-killed-annually-building-collisions-0. Accessed on July 15, 2017.

Pruetz, R. (2012, August/September). 'Lasting value', an excerpt from a recent APA Planners Press book that celebrates remarkable efforts to save rural areas and open space. *Planning, the Magazine of the American Planning Association,* 32–38.

Safarik, D., Ursini, S., & Wood, A. (2016). Megacities: Setting the scene. *CTBUH Journal, 4,* 30–39.

Sarrat, C., Lemonsu, A., Masson, V., & Guedalia, D. (2006). Impact of urban heat island on regional atmospheric pollution. *Atmospheric Environment, 40,* 1743–1758.

Wagner, M., & Viswanathan, V. (2016). Analyzing the impact of driving behavior at traffic lights on urban heat. *Procedia Engineering, 169,* 303–307.

Wells, W. (2014, February). Sweden, the green giant: A place where "sustainability" means collaboration. *Planning, the Magazine of the American Planning Association,* 42–47.

Wilson, J., Spinney, J., Millward, H., Scott, D., Hayden, A., & Tyedmers, P. (2013). Blame the exurbs, not the suburbs: Exploring the distribution of greenhouse gas emissions within a city region. *Energy Policy, 62,* 1329–1335.

Wood, A., & Salib, R. (2013). *Natural ventilation in high-rise office buildings.* Routledge, Taylor and Francis Group.

Worrel, G. (2012, January). Food groups: LA expands its menu of food policies and choices. *Planning, the Magazine of the American Planning Association,* 25–27.

Xia, J., Poon, D., & Mass, D. (2010). Case study: Shanghai tower. *CTBUH Journal, 2010,* 12–18.

Yeang, K. (2022). Designing sustainable tall buildings. In K. Al-Kodmany, P. Du, & M. M. Ali (Eds.), *Sustainable high-rise buildings: Innovation, design, and technology.* IET (Institution of Engineering and Technology) Publisher.

Chapter 2

Application of Expert Decision Systems for Optimal Fiber Selection for Green Building Design Components

Figen Balo and Lutfu S. Sua

Abstract

Composites based on fiber are commonly used in high-performance building materials. The composites mostly use petrochemically derived fibers like polyester and e-glass, due to their advantageous material features like high stiffness and strength. All the same, these fibers also have important short-comings related to energy consumption, recyclability, initial processing expense, resulting health hazards, and sustainability. Increasing environmental awareness and new sustainable building technologies are driving the research, development, and usage of "green" building materials, especially the development of biomaterials.

In this chapter, the natural fiber evaluation approach is applied, which covers a diverse set of criteria. Consequently, the comparative assessment of diverse natural fiber types is applied through the use of an expert decision system approach. The best performing fiber choice is made by comparatively evaluating the materials related to green building. The proposed fiber can be used and applied by green building material manufacturing companies in various countries or locations as a reference when selecting the fiber with the best performance.

Keywords: Natural fiber; synthetic fiber; AHP; biomass; renewable material; green building

1. Introduction

Since the 1960s, there has been an increasing demand for materials that are stronger and stiffer, yet lighter, for applications such as energy, civil, and aerospace. The composite products are new-generation products that have been

Pragmatic Engineering and Lifestyle, 21–37
Copyright © 2023 Figen Balo and Lutfu S. Sua
Published under exclusive licence by Emerald Publishing Limited
doi:10.1108/978-1-80262-997-220231002

designed to meet the needs of the industries' rapid technological changes. Composites are manufactured materials composed of two or more constituents with diverse chemical or physical features that stay separate and distinct within the finished construction. Composite products are noted for their combination of low weight and high structural stiffness. Composite materials are not new. Nature is brimming with examples of composite materials. Manufacturers may create features that meet the specifications for a specific function of a specific structure by using the right matrix content and reinforcement combination. Composite products provide efficiency that individual constituents cannot match, as well as the added value of design versatility. It is possible to create composite materials that meet the requirements of an electric motor, a boat, or an aircraft. These materials offer flexible design advantages. Composite products can be constituted to develop combinations of mechanical characteristics such as wear resistance, lubricity, strength, toughness, corrosion resistance, stiffness, damping, high-temperature performance, and esthetic features (Muthuraj et al., 2016).

Beneficial properties of natural fibers are that they are non-carcinogenic and biodegradable, with the added bonus of being cost-effective. Natural fibers are a valuable agricultural biomass that contribute to the world economy. Many different natural fibers have been developed for various applications. This diversity helps to keep an environmental balance in nature.

Engineers and scientists are having concerns fishing out novel resources of crude products that have comparable mechanical and physical features to synthetic fibers. Above all, natural fibers should come from sustainable sources. The main drawback for natural fibers is humidity adsorption; thus, it is necessary to modify the surface of the fibers through utilization of chemicals (Bongarde, 2014). There are numerous natural fibers that have the potential to be implemented for various applications: kenaf, pineapple, coir, sisal, abaca, cotton, bamboo, jute, palmyra, banana, talipot, flex, and hemp (Rowell et al., 1997; Schuh & Gayer, 1997; Yan et al., 2014). The primary differences between synthetic fibers and natural fibers are listed in Table 1.

Briefly, natural fibers have attracted attention in recent years due to their reasonably good properties, low cost, light weight, and minimal environmental effect (Muthuraj et al., 2016). For composite applications, there is a broad diversity of convenient natural fibers. Natural fibers are divided into groups based on the fiber source: bast, seed, leaf, and so on. Classification of some selected natural/plant fibers is given in Table 2.

Among composite materials, the most generally utilized ones are leaf fibers (palm, banana/abaca, and sisal), seed fibers (kapok, cotton, and coconut), and bast fibers (hemp, flax, kenaf, and jute). In general, much greater stiffnesses and strengths are achievable with plant-based natural fibers than with animal-based fibers (Shah et al., 2014). This makes vegetable-based fibers the most appropriate fibers for utilization in composite materials for structural needs. Additionally, natural fibers can be obtained in many countries around the world and can be extracted quickly. When comparing data from diverse studies, it should be noted that many variables (e.g., temperature and humidity) have an effect on fiber properties. In general, strength rises with increasing humidity and diminishes as

Table 1. Primary Differences Between Synthetic and Natural Fibers [14, 15].

Synthetic Fiber	Natural Fiber
They are not renewable (These fibers are easily prepared in laboratory or man-made. For example: aramid, e-glass, and the like)	They are renewable (Natural fibers are obtained in nature. For example: banana, cotton, and the like)
Synthetic fiber is not eco-friendly. For environment, some fibers are detrimental	Natural fiber is environment-friendly
Cost is high	Cost is low
Energy consumption is high	Energy consumption is low
There is not recyclability	There is recyclability
There is not CO_2 neutral	There is CO_2 neutral
There is wide distribution	There is wide distribution
There is health risk when inhaled	There is not health risk when inhaled
There is abrasion to the machine	There is not abrasion to the machine
Nonbiodegradable	Biodegradable
They do not have porosity as they are composed of chemical and thus do not act as adsorbents	They are well adsorbents and thus able to adsorb cold, heat, temperature, and so on relying on circumstances and nature of fibers
Not as comfortable as natural fibers	Comfortable in use
Used for filament fabrication, melting, dry or wet spinning processes	For filament fabrication, the spinning operation is not required.
Length of the fiber is controlled	Fiber's length is given in nature
Fiber widths can be controlled and the fibers can be modified to diverse configurations	Fiber length is found naturally and it is not prospective to modify the fiber's configuration
There is no doubt about its long or short fiber. It depends on person will.	Some short staple fibers found long staple fibers
It is obtained in filament shape but sometimes it could be transformed into cut or staple length	Fibers are obtained in filament or staple shape
For filament production, spinneret is essential	For spinning operation, no need for spinneret
For yarn production, chemical solution is essential	For yarn fabrication, no need for chemical resolution
No impurities or dust in synthetic fiber	In natural fiber, impurities or dust could exist

Table 1. *(Continued)*

Synthetic Fiber	Natural Fiber
Colors are added in resolution bath as needed	It grows with its natural color
It is easy to modify fiber configuration	It is not possible to modify fiber configuration
Crimp is applied on filament after passing spinneret	It contains natural crimp
Coloring is not so simple as natural fiber	It needs scouring and bleaching
Most of the times, they are hydrophobic	Most of the fibers are hydrophobic in nature
It becomes melted with burning	In most of the cases, it becomes ash after burning
Synthetic fiber is lower in worth than natural-based fiber	Comparatively, the natural-based fiber's price is greater than synthetic-based fiber's
After burning, chemical smell is found	After burning, the smell of the fiber is found as hair burn or paper burn
Synthetic fibers are utilized in multitasks	The natural fibers' use is restricted
Synthetic fibers are more durable than natural fibers	Comparatively less durable than synthetic fiber
High mechanical properties	Moderate mechanical properties
High special stiffness and strength	Lower stability than synthetic fibers
Low moisture sensitivity	High moisture sensitivity
Density is double high	Density is low
Low thermal sensitivity	High thermal sensitivity
High environmental production	Low environmental production
Limited resource	Infinite resource
Moderate environmental recyclability	Good environmental recyclability

temperature increases; the Young's modulus diminishes with humidity (Charlet et al., 2007). Because the fibers are typically stiffer and stronger than the matrix, the composite material's stiffness and strength usually increase with increasing fiber fraction (Beg, 2007).

Table 2. Classification of Some Selected Natural/Plant Fibers [16].

Natural Fibers/Plant Fibers

Leaf	Bast	Fruit	Seed	Wood	Grasses/Reed
Henequen	Flex	Luffa	Calotropis	Soft wood	Bagasse
Sisal	Jute	Coconut	Cotton	Hard wood	Bamboo
Curaua	Isora	Coir	Kapok		Oat
Manila	Hemp		Poplar		Wheat
Pineapple	Mesta				Rape
Yucca	Ramie				Rice
Palm	Kenaf				Rye
Piassava	Totora				Corn
Opuntia	Toina				Barley
Cabuja	Urena				Esparto
Screw pine					
Abaca					
Agaves					

A frequently unnoticed element of natural-fiber composites, porosity, has long been acknowledged to display a significant effect on mechanical features of composite materials. It originates due to entrapment of air during processing, restricted wettability of fibers, and voids within the fibers (Madsen et al., 2009). Porosity in natural-fiber composites has been shown to increase with increasing fiber fraction (Madsen & Lilholt, 2003).

It has been shown that fiber alignment improves the flexural and tensile properties of composites. Natural-fiber composites with aligned fibers are either fabricated manually, if the fiber is in the necessary orientation, or through alternative processing or textile infrastructure to fabricate a continual fiber shape. Relatively high strength values (101MPa with polylactide and 136MPa with epoxy) have been achieved with short fibers aligned using dynamic layer shaping. Fiber width has been shown to be significant for more arbitrarily aligned fiber composite materials, with those having longer fibers being applied better than those with shorter fiber. In general, the maximum flexural and tensile features for composites are accomplished with thermoset matrices for which the maximum values in decreasing order are for vinyl ester, epoxy, and then unsaturated polyester.

As mentioned above, humidity adsorption is one of the primary drawbacks with natural-fiber composites. Humidity adsorption increases with higher fiber fractions and temperature, in addition to being affected by fiber coupling/ treatment agent and fiber mixing. It is generally correlated with natural-fiber composites swelling and exhibiting decreased mechanical properties. Even with hydrophobic matrices like polypropylene, the flexural and tensile properties of natural-fiber composites have been shown to decrease substantially over several weeks immersed in water, with the reduction ratio rising with greater temperatures (Hargitai et al., 2008; Thwe & Liao, 2003). Maleic anhydride polypropylene utilized as a linkage substance has been observed to delay humidity absorption, decrease satiation humidity fraction, and improve composite properties during humidity testing of polypropylene-matrix composites (Beg & Pickering, 2008; Thwe & Liao, 2003). However, substantial degradation of properties, even with maleic anhydride polypropylene, are exhibited after 238 days, during which time satiation humidity fractions of 32, 45, and 59% occurred at 30, 50, and 70 °C, respectively.

Although the presence of natural fibers is usually thought to increase polymers' humidity adsorption, benefit of including natural-based fibers in artificial fiber supported thermoset composite materials has been obtained for prolonged exposure to humidity.

Green building components should be composed of materials that are sustainable, do not degrade the indoor air quality, reduce toxins and waste, reduce the amount of water used in their production, and are produced using renewable energy. Ecologically preferable materials are products that are, or can be, recycled and are natural-based products. Natural-based materials can be utilized as options for industrial products in structural applications, since they have many comparable physical and chemical properties as the synthetic products they substitute. Sometimes, they can even improve the product quality. Much progress and research has been accomplished during the last decades in the mechanical performance of natural-fiber composites. There have been improvements in fiber treatment, extraction, interfacial engineering, and selection.

This chapter reviews research on developing the stiffness, strength, toughness, weathering, and humidity resistance of natural-fiber composites. Natural-fiber composites are now comparable with synthetic-fiber composites for cost and stiffness; impact and tensile strengths are near those for synthetic-fiber composites. Applications of natural-fiber composites have grown significantly. However, additional research is needed to enhance their fire and humidity resistance.

Recently, to improve the stiffness and mechanical features of composites fabricated for green buildings, natural fibers have been used increasingly. This study deals with the favorability of the Analytical Hierarchy Process for choosing fiber materials for green building components. Furthermore, a few itemized case studies in selecting and evaluating natural-fiber and synthetic-fiber composites using expert decision systems are provided as a road map for the reader in forming, evaluating, and selecting appropriate natural-fiber composite materials for a specific application.

2. Analyzed Fibers

Yearly production of natural fibers (Faruk et al., 2012; John & Thomas, 2008; Taj et al., 2007) are summarized in Table 3 below.

Table 3. Yearly Production of Sources and Natural Fibers.

Fiber Source	Origin	World Production (10^3 Tons)
Sisal	Stem	380
Ramie	Stem	100
Pineapple	Leaf	Abundant
Piassava	Leaf	45
Palf		200
Oil palm	Fruit	Abundant
Nettle	Stem	Abundant
Kenaf	Stem	770
Jute	Stem	2,500
Isora	Stem	320
Henequen	Leaf	56
Hemp	Stem	215
Flax	Stem	810
Curaua	Leaf	60
Cotton	Stem	18,500
Coir	Stem	100
Banana	Fruit	200
Bamboo	Grass	30,000
Bagasse	Grass	764
B. mori silk	Stem	Abundant
Areca	Stem	10,000
Abaca	Stem	70

Climatic and geographic differences impact the properties of natural fibers. Thus, their properties are often reported as a range. The chemical composition of various natural fibers is displayed in Table 4.

The mechanical and chemical properties of natural fibers depend primarily upon the cellulose content, which ranges from fiber to fiber (Malkapuram et al., 2009). The natural fibers' cell structure and chemical composition are fairly complex (Siqueira et al., 2010). Chemical processes can increase the performance of natural fibers (John & Anandjiwala, 2007). Waikambo et al. found that the amorphous content, chemical composition, and crystalline packing order have an impact on a fiber's mechanical features (Waikambo, 2006). An individual fiber itself is a composite made of hemicelluloses, lignin, and cellulose, with small quantities of proteins, starch, sugars, and so on; it is a 3D biopolymer. The physical features of various natural fibers are provided in Table 5.

Table 4. Chemical Composition of Natural Fibers Analyzed.

Fiber	Hemicellulose (%)	Cellulose (%)	Pectin (%)	Lignin (%)	Wax (%)	Refs
Sisal	11.5	60	1.2	8	–	[35]
Ramie	14	72	2	0.8	–	[35]
Pineapple	17.5	80.5	4	8.3	–	[35]
Piassava	25.8	28.6	–	45	–	[36]
Palf	–	67.1–69.3	–	14.5–15.4	–	[37]
Oil palm	22–30	36–65	–	18–25	–	[38, 39]
Nettle	10	86	–	–	4	[36]
Kenaf	21	53.5	2	17	–	[35]
Jute	16	67	0.2	9	0.5	[35]
Isora	–	74	–	23	1.1	[35]
Henequen	28	60	–	8	0.5	[35]
Hemp	20	81	0.9	4	0.8	[35]
Flax	14.5	72.5	0.9	2.5	–	[35]
Curaua	5	73.6	–	7.5	–	[35]
Cotton	4	89	6	0.75	0.6	[35]
Coir	0.3	456	4	45	–	[35]
Banana	12.5	62.5	4	7.5	–	[40]
Bamboo	20.5	34.5	–	26	–	[35]
Bagasse	21	37	10	22	–	[35]
B. mori silk	27–38	31–45	–	14–19	2–7	[41]
Areca	13–15.42	57.35–8.21	–	23–24	0.12	[42]
Abaca	21	62.5	0.8	12	3	[35]

Synthetic fibers are identical throughout the world. They are made using standardized processes. Thus, their thermal and mechanical features do not change significantly. On the other hand, natural fibers have individual properties, since these fibers grow in natural conditions that are not identical all over the world, but differ from season to season and from region to region. These differences affect the plants' growth. Thus, their thermal and mechanical properties differ and comprehensive studies must be performed to analyze the natural fibers. Mechanical properties of natural and synthetic fibers (Khalil et al., 2012; Komuraiah et al., 2014) is provided in Table 6 below.

Table 5. The Physical Features of Natural Fibers Analyzed.

Fiber	Diameter, μm	Length, mm	Microfibrillar Angle, deg	Humidity Gain, %	References
Sisal	21	2.5	16	14	[50]
Ramie	31.55	160	–	8.5	[50]
Pineapple	50	–	10	–	[50]
Piassava	–	–	–	–	[51]
Palf	20–80	–	–	–	[52]
Oil palm	0.7–1.55	–	–	–	[52]
Nettle	–	–	–	14	[51]
Kenaf	19.8	2.35		17	[50]
Jute	18.4	2.55	8.1	17	[50]
Isora	1.3	–	–	1.2	[51, 53]
Henequen	1.2	–	–	–	[54]
Hemp	19.9	11.2	6.2	12	[50]
Flax	20	31.75	5	12	[50]
Curaua	1.4	–	–	–	[54]
Cotton	14.5	42	25	8.5	[50]
Coir	17.5	1.25	44	13	[50]
Banana	–	2.9	11	–	[50]
Bamboo	25	2	–	–	[50]
Bagasse	20	1.7	–	–	[50]
B. mori silk	–	–	–	–	[50]
Areca	114	–	–	–	[54]
Abaca	18.2	4.9	–	14	[50]

3. Methodology

Multicriteria decision-making methodologies are appropriate for assessing multiple options and selecting the optimum one. The stages of the analytic hierarchy process that could be utilized in decision-making problems are weight valuation, problem modeling, sensitivity analysis, and weight aggregation. As the starting point, main and sub criteria influencing the decision to be made are determined as well as their hierarchical relationships where applicable. In this study, all criteria are treated at the same level without any sub criteria involved. Then, cross-checks of the alternatives and criteria, and their comparative rankings, are computed.

The next step involves determining the weights of the criteria and subcriteria where available. This process is important to team decision-making as it presents

Table 6. Mechanical Features of Natural and Synthetic Fibers Analyzed.

Fiber	Young's Modulus (GPa)	Density (g/ cm3)	Tensile Strength (MPa)	Elongation (%)	Refs.
Carbon	230–240	1.4–1.75	4,000	1.4–1.8	[53]
Aramide	63–67	1.4–1.45	3,000–3,150	3.2–3.7	[53]
E-glass	73	2.5–2.55	2,000–3,500	2.5	[53]
S-glass	86.0	2.5	4,570	2.8	[59, 60]
Wood	40.0	1.5	1,000	–	[61]
Viscose	11.0	–	593	11.4	[61]
Twisted B.	3.82	–	156.27	20.57	[61]
Tussah silk	5.79	1.32	248.77	33.48	[61]
Spider silk	11–13	–	875–972	17–18	[62]
Sisal	9.4–22.0	1.5	511–635	2.0–2.5	[60, 63]
Ramie	61.4–128	1.5	400–938	3.6–3.8	[60, 63]
Pineapple	60–82	0.8–1.6	170–1,627	2.4	[62, 63]
Piassava	1.07–4.59	1.4	134–143	7.8–21.9	[53]
Palf	1.44–82.5	0.8–1.6	180–1,627	1.6–14.5	[53]
Oil palm	25.0	0.7–1.55	248	3.2	[63, 64]
Nettle	38.0	–	650	1.7	[64]
Kenaf	53.0	1.4	930	1.5	[63, 64]
Jute	26.5	1.3	393–773	1.5–1.8	[60, 63]
Isora	–	1.2–1.3	500–600	5–6	[53]
Henequen	10.1–16.3	1.2	430–570	3.7–5.9	[53]
Hemp	70.0	1.48	690	1.6	[60, 63]
Flax	27.6	1.5	345–1,035	2.7–3.2	[60, 63]
Curaua	11.8	1.4	500–1,150	3.7–4.3	[63]
Cotton	5.5–12.6	1.5–1.6	287–597	7.0–8.0	[60]
Coir	4.0–6.02	1.2	175	30.0	[60, 63]
Banana	27–32	–	529–914	3	[62]
Bamboo	11–17	0.6–1.1	140–230	–	[63
Bagasse	17.0	1.25	290	–	[63]
B. mori silk	6.10	1.33	208.45	19.55	[61]
Areca	1.12–3.15	0.7–0.8	147–322	10.2–13.15	[42]
Abaca	12.0	1.5	430–760	3–10	[63, 64]

Table 7. The Comparison Scale.

Intensity	Definition	Explanation
1	Equal significance	Two events take part evenly to the target
2	Slight or weak	
3	Moderate significance	Judgment and experience insignificantly favor one event over another
4	Moderate positive	
5	Strong significance	Judgment and experience firmly favor one event over another
6	Strong positive	
7	Too firm or indicated significance	One event is favored too firmly over another; its superiority indicated practically
8	Very, very strong	
9	Extreme significance	The proof favoring one event over another is of the maximum feasible order of assertion

significant factors for target accomplishment. To enable pairwise comparison of the criteria, a robust comparison scale needs to utilized such as the one provided in Table 7. As it can be observed from the table, "1" indicates an equal weight between to criteria while "9" indicates an extreme significance of one criterion compared to the other.

For the purpose of this study, eights factors are considered when evaluating a set of natural fibers and these criteria are compared based on the provided comparison scale. Table 8 provides the resulting decision matrix.

Through the pairwise comparison of the various chemical, physical, and mechanical features, the following weights are calculated based on expert evaluations. Twenty experts from the related industries and higher education institutions were asked to determine the relative weights of the predetermined criteria and the weights are calculated by taking the average values of the gathered data. The weights are presented in Fig. 1. As the figure indicates, the Young's modulus, which is a mechanical property of the fibers, is evaluated to be the most significant factor in evaluating the alternatives. If one looks at the mechanical properties of natural fibers, as an example, one finds different approaches to calculate tensile strength, normal distribution, *Weibull* distribution. When people determine tensile modulus, some of them use the stress–strain curve without calculating the system compliance. Similarly, there are many options to determine the fiber density yielding to different levels of error.

The final step of the methodology involves scoring the natural fiber alternatives for each criteria and multiplying the scores with the criteria weights. The resulting values are called *priority values*. These values are presented in Table 9. The last row of Table 9 shows the total priority values of each alternative. These priorities

Table 8. Decision Matrix for the Selected Criteria.

Matrix	Hemicellulose 1	Cellulose 2	Lignin 3	Diameter 4	Modulus 5	Density 6	Strength 7	Elongation 8	Weights
Hemicellulose 1	1	1/2	1/3	1/3	1/7	1/5	1/5	1/6	2.68%
Cellulose 2	2	1	1/2	1/2	1/5	1/3	1/3	1/4	4.36%
Lignin 3	3	2	1	1	1/4	1/2	1/3	1/5	6.20%
Diameter 4	3	2	1	1	1/6	1/4	1/5	1/6	5.18%
Modulus 5	7	5	4	6	1	4	3	2	32.00%
Density 6	5	3	2	4	1/4	1	1	1/2	12.68%
Strength 7	5	3	3	5	1/3	1	1	1/2	14.36%
Elongation 8	6	4	5	6	1/2	2	2	1	22.53%

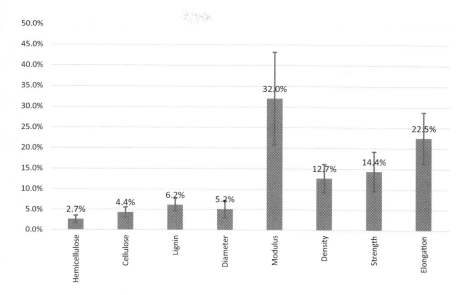

Fig. 1.   Weights of the Selected Criteria.

reflect on how much it contributes to the overall goal of finding the most optimum natural fiber alternative. The priorities along the hierarchy should add up to one. Since the values for the natural fiber characteristics have different units of measurement, they need to be normalized into a common scale. Thus, the initial values are divided by the sum of each row to find the normalized values. These values are then multiplied with the relative weights of each factor to obtain the priorities of the alternatives which constitutes Fig. 2.

Based on the total values, Fig. 2 exhibits the attractiveness of the selected natural fibers. Fig. 2 indicates that Ramie and Pineapple are the alternatives with the highest total scores, with values of 0.042 and 0.038, respectively. It can be observed that the relative attractiveness of Ramie and Pineapple mainly stems from their favorable modulus and elongation values. Based on the normalized eigenvector of the expert evaluations, relative weight of modulus is 0.320 while the weight of elongation is 0.225. These values ranked higher than the rest of the factors. Modulus value of Ramie ranges between 61.4 and 128 GPa and its elongation value is reported to be between 3.6 and 3.8%. These ranges for pineapple are 60–82 GPa and 2.4%, respectively.

4. Conclusions

Selecting the optimum fiber for a natural-fiber composite is a very complicated operation, as a comprehensive variety of alternatives and criteria have to be

Table 9. Priority Values of Natural Fiber Alternatives.

	Sisal	Ramie	Pineapple	Piassava	Palf	Oil Palm	Kenaf	Jute	Isora	Henequen	Hemp
Hemicellulose	0.001	0.001	0.001	0.002	0.001	0.002	0.002	0.001	0.001	0.002	0.001
Cellulose	0.002	0.002	0.002	0.001	0.002	0.001	0.001	0.002	0.002	0.002	0.002
Lignin	0.002	0.000	0.002	0.009	0.003	0.004	0.003	0.002	0.004	0.002	0.001
Diameter	0.002	0.003	0.005	0.002	0.005	0.000	0.002	0.002	0.000	0.000	0.002
Young's modulus	0.004	0.024	0.018	0.001	0.011	0.006	0.013	0.007	0.007	0.003	0.017
Density	0.012	0.012	0.010	0.012	0.010	0.009	0.012	0.011	0.010	0.010	0.012
Tensile strength	0.000	0.000	0.000	0.000	0.000	0.000	0.000	0.000	0.000	0.000	0.000
Elongation	0.000	0.000	0.000	0.000	0.000	0.000	0.000	0.000	0.000	0.000	0.000
Total	**0.022**	**0.042**	**0.038**	**0.026**	**0.031**	**0.023**	**0.033**	**0.024**	**0.025**	**0.019**	**0.036**

	Flax	Curaua	Cotton	Coir	Banana	Bamboo	Bagasse	B. Mori Silk	Areca	Abaca
Hemicellulose	0.001	0.000	0.000	0.000	0.001	0.002	0.002	0.002	0.001	0.002
Cellulose	0.002	0.002	0.002	0.012	0.002	0.001	0.001	0.001	0.001	0.002
Lignin	0.000	0.001	0.000	0.009	0.001	0.005	0.004	0.003	0.005	0.002
Diameter	0.002	0.000	0.001	0.002	0.002	0.002	0.002	0.002	0.011	0.002
Young's modulus	0.007	0.003	0.002	0.001	0.007	0.003	0.004	0.002	0.001	0.003
Density	0.012	0.012	0.013	0.010	0.011	0.007	0.010	0.011	0.006	0.012
Tensile strength	0.000	0.000	0.000	0.000	0.000	0.000	0.000	0.000	0.000	0.000
Elongation	0.000	0.000	0.000	0.000	0.000	0.000	0.000	0.000	0.000	0.000
Total	**0.025**	**0.018**	**0.019**	**0.034**	**0.025**	**0.020**	**0.023**	**0.022**	**0.025**	**0.023**

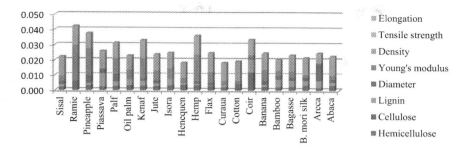

Fig. 2.   Total Priorities.

considered. Choosing the correct fiber composite materials is a multicriteria decision-making problem. Therefore, reliable and systematic mathematical models are required to evaluate a number of alternatives based on a variety of criteria. Appropriate mathematical models based on the expert decision systems for the fiber composite materials are able to aid decision-makers and designers obtain the optimal selection of such composite materials, with respect to their limitations and decision criteria.

Decision-support systems have been used in different areas of many industries; however, there is a research gap in using such systems for fiber-based green building components. To fill this gap, a multicriteria decision-making analysis method for the choice of the optimal fiber material to be utilized in green building components is addressed within the scope of this study. Based on the analysis conducted within this research, Ramie is found to be the optimum choice of natural fiber among the options with a priority of 0.042 while Curaua is determined to be the alternative with the lowest priority weight of 0018. This relative superiority of Ramie can be explained by its favorable position in terms of its modulus value which is the factor with the highest impact on the attractiveness of a natural fiber.

For natural fiber composites it has been difficult to make it to load bearing structural applications. Two main features prevent natural fibers from being used widely in structural application – they are not long continuous fibers and their properties vary according to soil type, weather condition, nutrients intake, etc. Contrary to glass, aramid or carbon fibers two distinct natural fibers will never have the same properties. They are used as short fibers (injection molded with a thermoplastic matrix) or nonwoven fibers.

When it comes to structural applications, one is interested in fiber density and tensile modulus. Density is more or less the same for various natural fibers. Decisive factor being therefore the modulus. Comparing various natural fiber types, hemp and flax always have the highest modulus regardless of the climate, soil type, etc. Modulus is determined by the microfibril angle.

Thermal conductivity/capacity is more or less the same for all the natural cellulosic materials.

The main contribution of this study is to propose a quantitative method which constitutes a basis on which a variety of factors on a multilevel hierarchical decision matrix can be included in the assessment of a number of alternatives.

Future direction of this research is to modify both the set of factors considered for the selection of the most optimum material and their relative priority values based on the expert reviews.

References

Beg, M. D. H. (2007). *The improvement of interfacial bonding, weathering and recycling of wood fibre reinforced polypropylene composites.* PhD thesis, University of Waikato.

Beg, M. D. H., & Pickering, K. L. (2008). Mechanical performance of Kraft fibre reinforced polypropylene composites: Influence of fibre length, fibre beating and hygrothermal ageing. *Composites Part A, 39*(11), 1748–1755.

Bongarde, V. D. S. U. S. (2014). Review on natural fiber reinforcement polymer composites. *International Journal of Engineering Science and Innovative Technology, 3*(2), 431–436.

Charlet, K., Baley, C., Morvan, C., Jernot, J. P., Gomina, M., & Breard, J. (2007). Characteristics of Hermes flax fibres as a function of their location in the stem and properties of the derived unidirectional composites. *Composites Part A, 38*(8), 1912–1921.

Faruk, O., Bledzki, A. K., Fink, H. P., & Sain, M. (2012). Biocomposites reinforced with natural fibers: 2000–2010. *Progress in Polymer Science, 37*(11), 1552–1596.

Hargitai, H., Rácz, I., & Anandjiwala, R. D. (2008). Development of hemp fiber reinforced polypropylene composites. *Journal of Thermoplastic Composite Materials, 21*(2), 165–174.

John, M. J., & Anandjiwala, R. D. (2007). Recent developments in chemical modification and characterization of natural fiber-reinforced composites. *Society of Plastic Engineers*, 187–198.

John, M. J., & Thomas, S. (2008). Biofibres and biocomposites. *Carbohydrate Polymers, 71*(3), 343–364.

Khalil, H. P. S. A., Jawaid, M., Hassan, A., Paridah, M. T., & Zaidon, A. (2012). Oil palm biomass fibres and recent advancement in oil palm biomass fibres based hybrid biocomposites. In N. Hu (Ed.), *Composites and their applications* (pp. 209–242). InTech.

Komuraiah, A., Kumar, N. S., & Prasad, B. D. (2014). Chemical composition of natural fibers and its influence on their mechanical properties. *Mechanics of Composite Materials, 50*, 359–376.

Madsen, B., & Lilholt, H. (2003). Physical and mechanical properties of unidirectional plant fibre composites – An evaluation of the influence of porosity. *Composites Science and Technology, 63*(9), 1265–1272.

Madsen, B., Thygesen, A., & Lilholt, H. (2009). Plant fibre composites – Porosity and stiffness. *Composites Science and Technology, 69*(7–8), 1057–1069.

Malkapuram, R., Kumar, V., & Ngi, Y. (2009). Recent development in natural fiber reinforced polypropylene composites. *Journal of Reinforced Plastics and Composites, 28*(10).

Muthuraj, R., Kumar, M., & Keerthiprasad, K. S. (2016). Characterization and comparison of natural and synthetic fiber composite laminates. *International Journal of Engineering and Techniques, 2*(5).

Rowell, R. M., Sanadi, A. R., Caulfield, D. F., & Jacobson, R. E. (1997). Utilization of natural fibres in plastic composites-problems and opportunities. *Lignocellulosic-Plastics Composites.*

Schuh, T., & Gayer, U. (1997). Automotive applications of natural fiber composites. Benefits for the environment and competitiveness with man-made materials. In A. L. Leao, F. X. Carvalho, & E. Frollini (Eds.), *Lignocellulosic-plastics composites.*

Shah, D. U., Porter, D., & Vollrath, F. (2014). Can silk become an effective reinforcing fibre? A property comparison with flax and glass reinforced composites. *Composites Science and Technology, 101*, 173–183.

Siqueira, G., Bras, J., & Dufresne, A. (2010). Cellulosic bionanocomposites: A review of preparation, properties, and applications. *Polymers, 2*, 728–765. https://doi.org/10.3390/polym2040728

Taj, S., Munawar, M. A., & Khan, S. U. (2007). Natural fiber-reinforced polymer composites. *Proceedings of the Pakistan Academy of Sciences, 44*(2), 129–144.

Thwe, M. M., & Liao, K. (2003). Environmental effects on bamboo-glass/polypropylene hybrid composites. *Journal of Material Science, 38*(2), 363–376.

Waikambo, M. (2006). Review of the history, properties and application of plant fibres. *African Journal of Science and Technology (AJST), Science and Engineering Services, 7*(2), 120–133.

Yan, L., Chouw, N., & Jayaraman, K. (2014). Flax fibre and its composites—A review. *Composites Part B: Engineering, 56*, 296–317.

Chapter 3

Advances in Solar-Driven Air-Conditioning Systems for Buildings

Sonali A. Deshmukh, Praveen Barmavatu, Mihir Kumar Das, Bukke Kiran Naik, Vineet Singh Sikarwar, Alety Shivakrishna, Radhamanohar Aepuru and Rathod Subash

Abstract

This study has covered many types of solar-powered air-conditioning systems that may be used as an alternative to traditional electrically powered air-conditioning systems in order to reduce energy usage. Solar adsorption air cooling is a great alternative to traditional vapor compression air-conditioning. Solar adsorption has several advantages over traditional vapor-compression systems, including being a green cooling technology which uses solar energy to drive the cycle, using pure water as an eco-friendly HFC-free refrigerant, and being mechanically simple with only the magnetic valves as moving parts. Several advancements and breakthroughs have been developed in the area of solar adsorption air-conditioners during the previous decade. However, further study is required before this technology can be put into practise. As a result, this book chapter highlights current research that adds to the understanding of solar adsorption air-conditioning technologies, with a focus on practical research. These systems have the potential to become the next iteration of air-conditioning systems, with the benefit of lowering energy usage while using plentiful solar energy supplies to supply the cooling demand.

Keywords: Solar-driven air-conditioning systems; solar adsorption air-conditioning system; green HFC-free refrigerant; vapor compression system; cooling technologies; solar-powered heating systems

Pragmatic Engineering and Lifestyle, 39–60
doi:10.1108/978-1-80262-997-220231003

1. Need and Demand of Air-Conditioning Systems

Saidur (2009) is interested in estimating energy usage in Malaysian office buildings as well as the energy utilization of important equipment. Malaysia's energy intensity (EI) – a measure of a building's energy efficiency – is calculated and compared to a number of other countries. Air-conditioners are the largest energy consumers in office buildings (57%), followed by lights (19%), elevators and pumps (18%), and miscellaneous equipment (6%). In tropical climes, air-conditioning (AC) systems use the bulk of the energy consumed by buildings. Office buildings consume roughly 70–300 kWh/m^2 per year, which is 10–20 times more than residential structures. Human comfort is becoming increasingly popular. According to the International Center of Refrigeration in Paris, approximately 15% of all electricity produced in the world is used for refrigeration and air-conditioning practices of different kinds, with air-conditioning systems accounting for 45% of total energy consumption in households and commercial buildings. The majority of this need is fulfilled by refrigeration systems based on vapor compression. Some vapor absorption-based refrigeration systems have recently become available for industrial and commercial building application at a low cost. Globalization, the improvement of living circumstances in developing regions, and the expansion of communication networks encourage developed-country lifestyles and lead to consumption habits that, without a doubt, will deplete fossil fuel supplies and have a major environmental impact. Hundreds of studies on the development of energy-efficient thermally driven adsorption cooling systems have been published in the literature. Finding the best adsorbent–adsorbate pair for a typical adsorption cooling application is usually difficult. The physical structure of the adsorbent, adsorption equilibrium, and adsorbent–adsorbate interactions are all important factors in the performance of a working pair. As a result, (Faizan et al., 2020) intend to offer a thorough study and comparison using a variety of adsorbent–adsorbate pairings. Adsorption isotherms are used to identify and compare data on adsorption uptake. The variables of adsorption isotherms models are provided and evaluated as a whole. Wajid et al. (2021) study that more research is needed to get this technology to a practical level. If a solar-powered environmental management system was employed to meet both the heating and cooling needs of the building it served, it would be more cost-effective. Various solar-powered heating systems have been thoroughly tested, but solar-powered air-conditioners have gotten nothing more than a passing glance. Solar collectors can be used to supply energy for absorption cooling, desiccant cooling, and Rankine-vapor compression cycles, among other things. Hybrid solar cooling systems are also an option. Despite a considerable potential market, present solar cooling systems are not cost competitive with electricity- or gas-fired air-conditioning systems due to their high initial costs as explained by (Yasiri et al., 2022). The cost of solar cooling systems might be reduced by lowering component costs and enhancing component performance. Solar component costs will be lowered as a result of improvements such as reduced collector area and reduced collector cost due to enhanced system performance. Several solar-powered refrigeration systems have been suggested and are in the works, including sorption systems for liquid/vapor, solid/vapor, adsorption, vapor compression, and photovoltaic-vapor compression. According to (Li & Sumathy, 2000), most of the abovementioned methods have not been financially justified. Thus, solar energy may be used for air-conditioning in two ways: first, through solar photovoltaic cells, and later, through vapor-compression cycles and heat-driven sorption systems. Solar photovoltaic cell efficiency is improving slowly; therefore the initial cost is still relatively high. Vapor absorption systems are already commercially available among the heat-driven systems; however, they typically have a capacity of more than 30 TR. For lesser capacities, they have constraints. Adsorption-based systems might fill this need by replacing high-energy, fossil-fuel-consuming, and environmentally harmful vapor-compression devices. Even if half of the current market for small air-conditioning systems can also be replaced by a solar-powered adsorption system, a significant quantity of electrical energy may be saved and a significant amount of carbon credit obtained as explained by (Robbins & Garimella, 2020). The adsorption system's natural refrigerants have no potential to deplete the ozone layer or cause global warming. Adsorption systems are small and quiet, and they are less susceptible to shocks and placement. Adsorbent does not need to be replaced frequently. They have a small number of rotating parts, no refrigerant/adsorbent pump, and very minor maintenance and servicing difficulties. In contrast to absorbent systems, there are no corrosion or crystallization issues. For part-load operation, regeneration temperature flexibility is far greater than in absorption systems. A lot of research is being done on adsorbent–adsorbate couples such zeolite–water, activated carbon–ammonia, activated carbon–methanol, and silica gel–water, among others. Due to its low regeneration temperature, the silica gel–water combination is perfect for solar energy use. Simple flat plate or evacuated tube solar collector systems cannot achieve these temperatures. Although the activated carbon–methanol duo operates at low regeneration temperatures, it is more suited to ice creation and freezing. Water is a better choice for air-conditioning since it has a greater latent heat of vaporization and can produce chilled water temperatures of 8–10°C. Researchers have made a variety of attempts to increase the performance of solar applied air-conditioning (chiller) subsystems. The most significant parameter in the design and manufacturing of a solar-powered air-conditioning system is the generator inlet temperature of the chiller. Other influencing elements for system functioning include collector selection, system design, and configuration (Li & Sumathy, 2000).

1.1 Disadvantages of Conventional Air-Conditioning Systems

Traditional vapor-compression machines utilize a significant amount of electrical energy, resulting in the depletion of valuable fossil fuel resources as well as the emission of several greenhouse gases. The majority of refrigerants in use deplete the ozone layer, with some emitting significant amounts of greenhouse gases. CFCs, HCFCs, and HFCs, which deplete the ozone layer and produce greenhouse gases, are banned by the Montreal and Kyoto Protocols. As a result, scientists working in the sector are paying close attention to the use of thermal energy, such as solar energy, for air-conditioning and refrigeration (Faizan et al., 2020).

2. Desiccant Technologies

Lisa and Stephen (2012) examine how desiccant technology has become a vital weapon in the industry's arsenal of space conditioning choices. Desiccant cooling units have benefits over vapor phase and absorption units in some cooling applications. Desiccant units, for example, do not use ozone-depleting refrigerants and may operate on natural gas, solar thermal energy, or waste heat, reducing peak electric demand. Advanced desiccant materials with higher sorption capacity, favorable equilibrium isotherms, and superior moisture and heat rates have been the focus of recent studies in the development of a solar desiccant cooling system. Improved performance will cut these systems' beginning costs, making them a more appealing alternative to conventional vapor compression systems. Solar solid and liquid desiccants are still employed in air-conditioning to enhance indoor air quality since they are also ecologically beneficial. During the night, sensible cooling is given, and the desiccant is renewed by solar heat during the day. The system has two modes of operation: adsorption and daylight desorption. Air is routed to the desiccant bed on the roof during the night to eliminate moisture. The sorption heat is released into the atmosphere. After passing through the desiccant bed, the air is cooled further by passing through an evaporative cooler to enhance the humidity level. This air is then blown into the conditioned zone, where it absorbs both heat and moisture from the surrounding environment. Adsorption or absorption describes the process of attracting and keeping moisture, depending on whether the desiccant undergoes a chemical change as it absorbs moisture. Adsorption changes the desiccant only by adding the weight of vapors, which is analogous in some respects to a sponge soaking up water, as examined by (Zhai & Wang, 2009). Desiccants are a kind of sorbent that has a strong attraction to water. As it needs low temperature heat (50–90°C) and enables for high density lossless energy storage in the form of concentrated desiccant, liquid desiccant cooling is particularly well suited to solar applications. The capacity to store energy is a major asset for solar applications when comparing liquid desiccant systems to solid desiccant or rotary wheel dehumidifiers. Dedicated Outdoor Air Systems, which precondition ventilation air and provide better control over delivery air humidity, are common configurations for LDAC systems. Additional sensible cooling can be provided by evaporative coolers or undersized vapor compression chillers. To condition the process air to the necessary comfort conditions, solar-assisted desiccant cooling comprises the following basic components.

(1) Continuously revolving rotary wheel loaded with a nominal silica gel matrix between both the process and regeneration air streams.

(2) Rotating between the process and regeneration air streams is a sensible heat wheel (rotary regenerator) (the wheel transfers heat from the process side to the regeneration side).

(3) Evaporative air coolers on the process and regeneration sides, with a solar collector storage subsystem to provide the needed thermal energy for regeneration.

(4) As a backup for the solar subsystem, a gas-fired auxiliary heater is used.

(5) Two thermostats, one to activate the desiccant system and the other to activate the vapor-compression system Hybrid systems, which combine desiccant dehumidifiers with traditional cooling systems, have been shown to save significant amounts of energy. If the rotating dehumidifier matrix is appropriately designed and appropriate selections for the kind and amount of desiccant are made, the required energy for regeneration can be provided by low temperature heat sources.

2.1 System Performance of LDAC Systems Are Measured

The thermal coefficient of performance (COP) is used to describe thermally driven cooling systems as explained by (Ayadi & Al-Dahidi, 2019). It indicates the ratio of cooling energy to heat input during a given time period. The electrical COP, on the other hand, is the ratio of cooling energy to the electrical input energy required to run pumps and fans as explained by (Wajid et al., 2021). The system's capacity (kW) is also significant since it must be able to handle the cooling demand for the building in which it is installed. The solar collector efficiency and solar fraction are used to calculate the performance of a solar cooling system. The collector efficiency is defined as the ratio of collected energy to incoming solar energy, and it may be calculated using total area or absorber area. The solar fraction, which may be computed using Eq. (1), indicates what proportion of the load is met by solar energy.

$$SF = 1 - \frac{Q_{aux}}{Q_{load}} \tag{1}$$

3. Solar Air-Conditioning/Cooling Systems

The solar adsorption air-conditioning system (SADCS) is a great replacement for the traditional vapor compression system (VCS). SADCS has several benefits over VCS systems, including being a green cooling technology that uses solar energy to drive the adsorption/desorption cycle, using pure water as a green HFC-free refrigerant, and being mechanically simple with only the magnetic valves as moving components. Several advances and discoveries in the field of SADCS research have been made in the recent decade.

3.1 Solar Absorption Systems

A typical absorption refrigeration cycle is carried out by solar absorption refrigeration systems. Its energy source is solar light as used by (Akhtar et al., 2015; Ramli et al., 2019). Absorption systems often employ evacuated tube collectors (ETCs) since they require a minimum temperature of 80°C. Solar concentrating collectors, as illustrated in Fig. 1, can be used to increase the overall system's efficiency.

3.2 Solar Adsorption Systems

Solar-powered adsorption refrigeration systems function in such a manner that the secondary fluid alternately feeds energy to one adsorber and then to another, as illustrated in the diagram. The refrigerant emitted from the bed during this procedure cools the room by passing via a condenser, expansion valve, and evaporator. Heat is transmitted from the condenser to the cooling tower. A varied temperature of solar fluid is required to thermally drive the adsorption cycle depending on the kind of working pairs (solid and fluid) (Lattieff et al., 2019; Zhu et al., 2018). Even flat plate solar collectors operating at a temperature of 80–90°C can be used if sun irradiation is high (Fig. 2).

3.3 Existing Solar Air-Conditioning Systems

Bilgili (2011) investigated the efficacy of a SEVCR system installed in a building in southern Turkey on 23 August in the cooling season, which runs from May to September. The hourly fluctuations of several parameters, including coefficient of

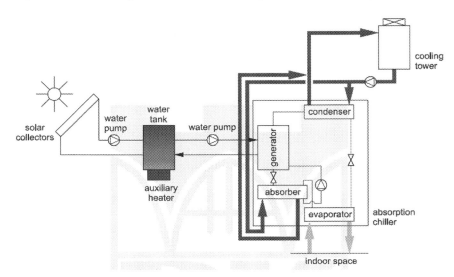

Fig. 1.   Solar Absorption Refrigeration (Dorota et al., 2014).

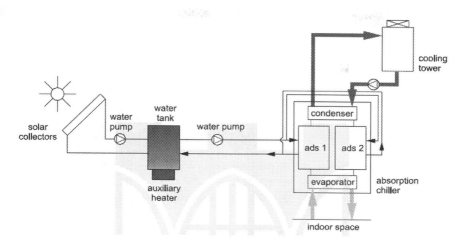

Fig. 2.　Solar Adsorption Refrigeration (Dorota et al., 2014).

performance, condenser capacity, and compressor power consumption, were then determined. In addition, based on hourly average solar radiation data, the minimal photovoltaic panel surface area to fulfill the compressor power need was estimated. At 15:00 p.m. on August 23, the greatest compressor power usage for evaporating temperature $T_e = 0°C$ was 2.53 kW. The needed surface area of photovoltaic panels was calculated to be roughly 31.26 square meters. In the southern part of Turkey, it was discovered that the SE-VCR system may be utilized for daytime home/office cooling. The experimental setup is depicted schematically in Fig. 3.

Almohammadi and Harby (2020) use a multiple objective genetic algorithm (MOGA) that combines a Kriging-based response surface to optimize the operational parameters. Hot, cooling, and chilled water temperatures and mass flow rates are among the eight operational factors examined, as are cycle and switching periods. Three axial finned tubes heat exchangers linked in parallel have been constructed and tested in an innovative SDACS. To anticipate system performance, a nonequilibrium lumped parameter model was created. The results of the optimization method and simulation are compared to those obtained experimentally, and there are good agreements with a maximum error of 10%. At rated operating circumstances, the suggested SDACS may provide around 0.56 kW cooling power with a COP of about 0.52. When compared to the rated operating circumstances at the same design parameters, the optimized operating conditions employing MOGA enhance the SCP by 51.7% and the system COP by 21%. Schematic diagram for proposed SDACS system is shown in Fig. 4.

The system design specifications are

- Silica gel/water is used as adsorbent/adsorbate pair
- Collector type ETSC

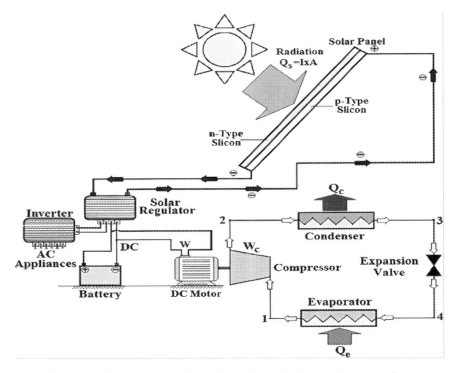

Fig. 3.   Components of the Solar Electric-Vapor Compression Refrigeration (SE-VCR) System. *Source:* Reprinted from Bilgili (2011), Copyright 2011, with permission from Elsevier.

- SADS cooling effect (capacity)
- 220 kW/kg of Silica gel
- COP is 0.63
- It is observed that the best performance is at 85°C driving temperature, and above this limit, the improvement is marginal.

Fig. 5 is used to determine the best operating settings for a solar-powered adsorption cooling system in order to maximize thermal performance. A suggested SDACS was created and built for this purpose to study system performance. To improve the operating conditions of the SDACS and hence system output, a multiobjective genetic algorithm (MOGA) using a Kriging-based response surface is used.

The system design specifications developed by (Sha & Baiju, 2020) are as given below

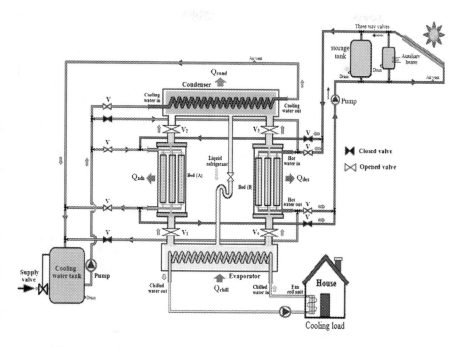

Fig. 4. Schematic Diagram of the Proposed SDACS. *Source:* Reprinted from *Energy*, 206, Mehmet Bilgili, "Operational conditions optimization of a proposed solar-powered adsorption cooling system: Experimental, modeling, and optimization algorithm techniques," Copyright 2020, with permission from Elsevier.

- Activated carbon/ethanol is used as adsorbent/adsorbate pair
- Collector type PTSC
- SADS cooling effect (capacity)
- 500 W
- COP is 0.68
- It is observed that activated carbon/ethanol is the best pair for the continuous SADS cooling cycle

Jribi et al. (2014) investigated a transient mathematical model of a four-bed adsorption chiller with Maxsorb III as the adsorbent and CO_2 as the refrigerant. For varying heating and cooling water intake temperatures, the cyclic steady-state system's performance is shown. Due to the low critical point of CO_2 ($T_c = 31°C$), it is discovered that desorption pressure has a significant impact on performance. At a driving heat source temperature of 95°C, a cooling temperature of 27°C, and an optimal desorption pressure of 79 bar, the CO_2-based adsorption chiller

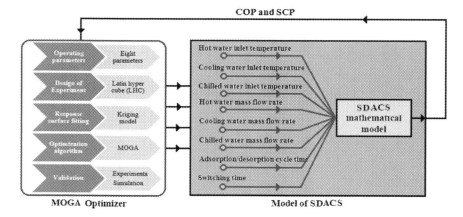

Fig. 5.   Optimization Process for the Investigated Eight Operating Conditions for SDACS. *Source:* Reprinted from Almohammadi and Harby (2020), Copyright 2020, with permission from Elsevier.

generates 2 kW of cooling power and has a COP of 0.1. The current thermal compression air-conditioning system is suited for both home and mobile air-conditioning applications since it may be powered by solar energy or waste heat from internal combustion engines as shown in Fig. 6.

The system design specifications are

- Activated carbon (Maxsorb III)/CO_2 is used as adsorbent/adsorbate pair
- Collector type
- SADS cooling effect (capacity)
- 2 KW
- COP is 0.1
- It is observed that heat and mass recovery techniques are necessary to improve the system performance.

Chen et al. (2018) examined the efficacy of an adsorption refrigeration system employing SAPO-34 zeolite and water as the working pair under sun heating conditions. The experiment revealed the effects of cycle time on system performance as well as the dynamic change of temperature and pressure in the adsorption bed. The best cycle time of the system was determined by analyzing the link between solar energy intake and cooling output. As the system was assessed by the whole cycle time, it was discovered that both the productivity coefficient COP and the specified cooling power SCP displayed a maximum value with regard to the adsorption time However, for their highest values, the COP and the SCP did not have the same adsorption time. The ability of the bed to lower its adsorption rate over time is thought to be crucial for determining the system's

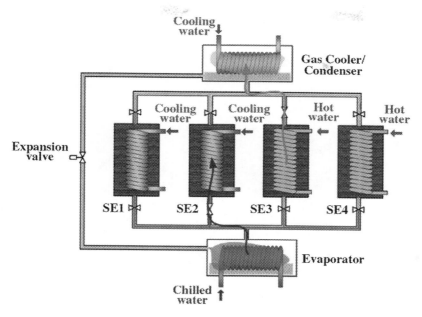

Fig. 6.   Schematic Diagram of the Four-Bed-adsorption Cooling System.
SE1 (sorption element 1), SE2, SE3, and SE4 are in pre-cooling, adsorption,
desorption, and preheating processes, respectively. *Source:* Reprinted from Jribi
et al. (2014), Copyright 2014, with permission from Elsevier.

ideal cycle time. The very extended adsorption period has no effect on the system's cooling performance.

3.4 Effect of Adsorption Time on the Refrigeration Performance

As previously stated, the COP measures the system's ability to convert solar energy into refrigeration capacity. The COP gradually rose over time as shown in Fig. 7. It meant that the longer the adsorption period, the higher the system's performance.

3.5 Effect of the Adsorption Time on SCP

As a result, the specific refrigeration power SCP1, which was determined by the adsorption time, dropped monotonically with the adsorption time. The change in SCP2 that was determined by the entire cycle duration, on the other hand, was substantially different. There was a maximum value for the SCP2 as well as the system's performance coefficient COP when the adsorption time changed. Although the ideal adsorption times for the COP and SCP2 were not the same,

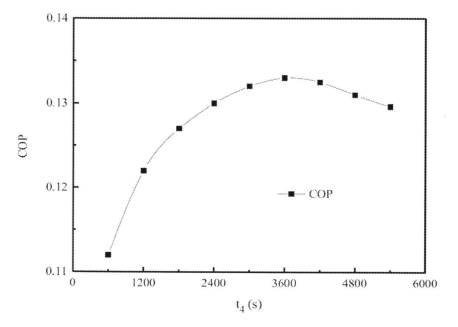

Fig. 7.   Variation of the COP With the Adsorption Time.

they were only around 1200s apart. This verification has given credence to the solar adsorption refrigeration system's optimization potential.

The system design specifications are

- SAPO-34 zeolite/water is used as adsorbent/adsorbate pair
- Collector type PTSC
- SADS cooling effect (capacity) 181.3–298.9 kJ
- COP is 0.112–0.13

The adsorption capacity changes nonlinearly with the adsorption time, it is seen. The ideal period for adsorption was not the same for the system COP and the particular cooling power. Alahmer et al. (2020) evaluate the performance of a two-bed silica gel–water adsorption refrigeration system powered by solar thermal energy under climatic conditions typical of Perth, Australia. Based on actual data from Standard forms software, version 7.0 for Perth, Australia, a Fourier series is used to approximate solar radiation. Payback Period and Life-Cycle Saving are two economic approaches used to analyze system economics and optimize the requirement for solar collector areas. The order of the Fourier series had no major influence on the simulated radiation data, and a three-order Fourier series was excellent enough to represent the actual solar radiation, according to the research. On a normal summer day, the chiller's average cooling capacity at peak hour (13:00) is approximately 11 kW, and the cycle chiller system and solar system COP are around 0.5 and 0.3, respectively. The payback period for the solar adsorption system tested was about 11 years, and the best solar collector area was around 38 m² if a compound parabolic collector (CPC) panel was employed, according to the economic study. The study found that using solar-driven adsorption cooling for weather circumstances similar to those in Perth, Australia, is both economically and technically feasible.

A schematic illustration of an adsorption cooling system using CPC solar collectors is shown in Fig. 8. A traditional two-bed adsorption chiller (with one adsorber/desorber pair, an evaporator, a condenser, and four refrigerant valves), solar thermal system CPCs, metallic tubes for cold and hot working fluids, and a cooling tower make up the majority of the system. The "evaporator-adsorption process" occurs when water evaporates from the evaporator and is absorbed by the adsorber through valve V1. At the same time, the "desorption-condensation process" sends desorbed water vapor to the condenser through valve V4.

The system design specifications are

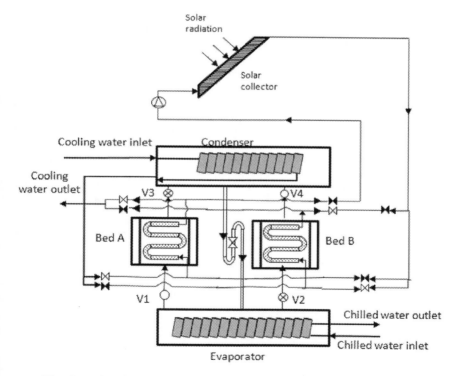

Fig. 8.   A Schematic Drawing of an Adsorption Cooling Chiller Driven by the Compound Parabolic Collector (CPC) Solar Collector Panel.

- Silica gel/water is used as adsorbent/adsorbate pair
- Collector type PTSC
- SADS cooling effect (capacity) 11 kW
- COP is 0.5
- In the afternoon between 10:00 and 17:00, COPs in the range of 0.4–0.55 were measured, indicating efficient cooling for commercial or residential structures.

The system design specifications are

- Activated carbon/methanol is used as adsorbent/adsorbate pair
- Collector type ETSC
- SADS cooling effect (capacity)
- 35 kW
- COP is 0.403

When employing a 20 kW and 40 kW auxiliary heater, the collector area was lowered by 63% and 33%, respectively. The solar percent rises as the solar loop mass flow rate rises. The solar portion was not affected by a greater flow rate (≥ 0.2 kg/s). Solar adsorption cooling, as a green energy usage technique, is a viable approach to utilize solar energy in the future by (Roumpedakis et al., 2020). The performance of a solar adsorption cooling system that used silica gel–water as the working pair was assessed experimentally in order to increase the efficiency of such a system. The influence of adsorption duration on system performance was primarily examined in the study, which used a mono-axial solar collector with a tracking parabolic trough to give heat to the adsorption bed. The system's performance was at its peak when the adsorption period was 45 minutes, according to the data. As a result, the system's performance coefficient increased to 0.258. Liu et al. (2021) also did a comparison of the optimal performance of the silica gel system to the SAPO-34 zeolite system. The cooling capacity and performance coefficient of the silica gel system were found to be more dependent on the adsorption time than those of the SAPO-34 zeolite system. The coefficient of performance of the silica gel system was 1.93 times that of the SAPO-34 zeolite system at the optimal adsorption time. In general, as compared to the SAPO-34 zeolite-water pair employed in the solar adsorption cooling system, the silica gel–water pair has performed better.

The system design specifications for a system are as shown in Fig. 9.

- Silica gel/water zeolite (SAPO-34)/water is used as adsorbent/adsorbate pair
- Collector type Tracked PTSC
- SADS cooling effect (capacity) 548.8 kJ 284.2 kJ
- COP is 0.258 0.133
- It is observed that silica gel showed better performance than the zeolite-based on the COP and cooling power. The zeolite-based SADS was less sensitive to the adsorption time changing as compared to the silica gel-based SADS.

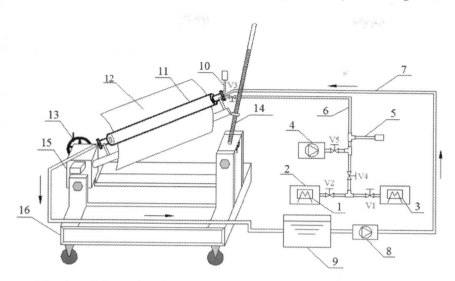

Fig. 9.   Schematic Diagram of the Solar Adsorption Refrigeration System [1-Evaporator; 2-Insulation Tank; 3-Condenser; 4-Vacuum Pump; 5-Pressure Sensor of Evaporator and Condenser; 6-Steam Pipe; 7-Cooling Water Pipe; 8-Circulating Water Pump; 9-Water Tank; 10-Pressure Sensor of Adsorption Bed; 11-Adsorption Bed; 12-Parabolic Trough Collector; 13-Adjusting Handle for the Trough; 14-Adjusting Lever for the Trough; 15-Stepper Motor; 16-Support Frame]. *Source:* Reprinted from Liu et al. (2021), Copyright 2021, with permission from Elsevier.

Tso et al. (2015) designed an adsorption cooling system utilizing a unique composite material (zeolite 13X/CaCl$_2$) as the adsorbent and water as the adsorbate, and examined the system performance experimentally under various operating circumstances. When compared to the zeolite 13X adsorbent, the composite adsorbent may be desorbed at a substantially lower temperature. The SCP of the adsorption cooling system employing the composite adsorbent is about 30% higher than that of the same system utilizing silica gel as the adsorbent under the same operating conditions. Although the adsorption cooling system with the composite adsorbent requires a longer adsorption/desorption phase time (cycle time), it provides superior cooling performance than the silica gel adsorbent when a lower chilled water temperature is required. The adsorption cooling system's various working sequences (i.e., heat recovery, mass recovery, preheating and precooling cycles) have also been explored. The heat and mass recovery cycle improves the SCP and COP of the adsorption cooling system by about 126% and 125%, respectively, over the adsorption cooling system. Heat recovery, on the other hand, necessitates the addition of additional equipment to the adsorption cooling system. As a result, mass recovery along with the preheating and precooling cycle is preferable, resulting in SCP and COP values of around 106 W/kg and 0.16, respectively. When compared to the fundamental cycle, the rise is between 129 and 100% for a system which is shown in Fig. 10.

Fig. 11 shows effect of desorption temperature on SCP and COP for 31°C cooling water inlet temperature; 15 min adsorption/desorption phase time for both adsorbents; 8 kg/min hot and cooling water mass flow rate; 3.6 kg/min chilled water mass flow rate; 14°C chilled water inlet temperature; 50 s heat recovery time and 10 s mass recovery time].

The system design specifications are

- Zeolite 13X/CaCl$_2$)/water is used as adsorbent/adsorbate pair
- Collector type PTSC
- SADS cooling effect (capacity) 371 W
- COP is 0.16

Fig. 10.   A Prototype of the Adsorption Cooling System. [Remarks: 1: cooling water tank; 2: isothermal water circulator; 3: chilled water tank (evaporator); 4: adsorber A; 5: adsorber B; 6: hot water tank; 7: condenser; 8: control system]. *Source:* Reprinted from Tso et al. (2015), Copyright 2015, with permission from Elsevier.

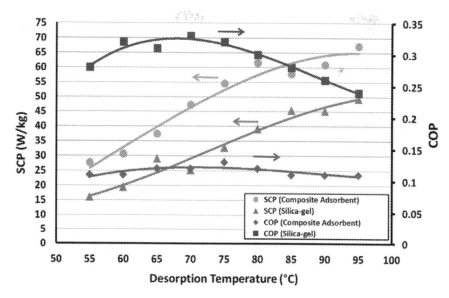

Fig. 11.   Effect of Desorption Temperature on the SCP and COP.
Source: Reprinted from Tso et al. (2015), Copyright 2015, with permission from Elsevier.

- It has been discovered that the heat and mass recovery cycle improves the COP of the SADS by around 125%. When compared to the basic cycle based zeolite 13X/water, the mass recovery, along with the preheating and precooling cycles, resulted in a 100% improvement in COP.

A zeolite–water adsorption chiller is coupled with solar thermal collectors in the proposed system. The designed sorption chiller has a cooling capability of more than 10 kW when SAPO-34 is used as the adsorbent. A backup electrically driven heat pump is linked with the adsorption chiller to lower the chiller's capacity and hence the needed solar field area while simultaneously improving the efficiency of part-load operation. The backup heat pump has a cooling capability of 10 kW and is mostly utilized to handle peak demands. Three rows of sophisticated evacuated tube collectors with a total surface area of 40 m² make up the solar field. A 1 m³ heat storage tank is also included in the system, which absorbs solar energy surges and allows the sorption module to operate more consistently. The heat rejection unit for both the adsorption chiller and the backup heat pump, retrofitted for the specific use, is a "V-shaped" dry cooler. Fig. 12 provide an overview of the prototype, including the installed measurement instruments. A propylene glycol solution is employed as the working medium for the solar subsystem to improve performance and allow risk-free operation at low ambient temperatures. Furthermore, the adsorption chiller's secondary circuits use only

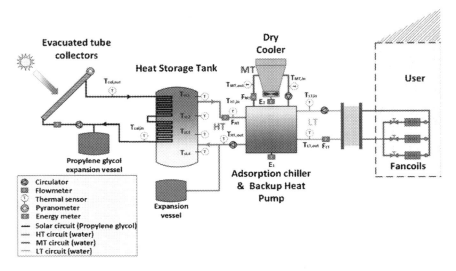

Fig. 12. Schematic of System Prototype With All the Involved Measuring Devices.

clean water. The heat coils in the 1 m³ heat storage tank transmit heat from the glycol solution to hot water, which powers the adsorption chiller. The severe environmental impact of traditional cooling and heating systems has raised the demand for renewable energy deployment to meet thermal load requirements. To that end, the suggested system combines a zeolite–water adsorption chiller with a traditional vapor compression refrigeration system to provide an efficient solar-powered alternative. The system is built to run on an intermittent heat supply of low-temperature solar thermal energy (<90°C) from evacuated tube collectors. In cooling mode operation, a prototype was created and tested. The results of individual component testing revealed that the adsorption chiller was performing well, with a maximum power coefficient (COP) of 0.65. The greatest recorded COP for the system's combined performance, measured during a normal summer week in Athens, was roughly 0.575, owing to lower driving temperatures in the region of 75°C. 5.8 was the associated mean energy efficiency ratio (EER).

The system design specifications are

- Zeolite (SAPO-34)/water is used as adsorbent/adsorbate pair
- Collector type ETSC
- SADS cooling effect (capacity) 10 kW
- COP is 0.575
- It is observed that the best performance of the SADS was reported at 75°C, in which the cooling load of the building was satisfactorily met with a mean energy efficiency ratio (EER) of 5.8.

Due to the usage of the sun's diurnal cycle for system management, attempts to construct autonomous solar adsorption chillers without additional controls to govern system operation have generated inadequate cooling capacity and poor COPs. This chapter investigates a thermally driven solar adsorption chiller with constant cooling rates. The working pair activated carbon/ammonia is examined in a model for a single-bed setup. The calculated heat input is equivalent to 1 m² of collector area. Instead of relying on the solar cycle, autonomous operation is achieved by applying innovative thermally activated controls to manage the heating and cooling of the adsorbent bed. During the adsorption phase, heat from the collector is stored in a hot-side thermal mass, and cooling is provided by an air-cooled heat sink. For operation, the system does not require any pumping power or auxiliary equipment. This system is appropriate for cooling applications in undeveloped countries or in installations that are not linked to the grid due to its basic design and lack of pricey complicated components.

To regulate the system without active controls, the method examined here uses thermal switching and the change in mass of the adsorbent bed. Furthermore, using separate heating and cooling structures that are not in touch with the bed decreases the amount of thermal mass that needs to be heated and cooled in each cycle, improving system performance. The mechanism is depicted in Fig. 13. As sunlight is absorbed by the solar panel, it is transferred to water via a heat pipe. The thermal switch is connected to the heated water. The water delivers heat to the adsorption bed when the switch is turned on. When the switch is turned off, the water serves as a thermal reservoir, storing the energy collected by the solar collectors. A set of contact pads on the heat pipe and on the adsorbent bed make up the switch. When the adsorbent absorbs a certain quantity of refrigerant, the bed shifts and the two sets of pads come into contact. As this is the primary thermal barrier, heat flow increases, and the temperature of the adsorbent bed rises.

The system design specifications are

- Activated carbon–ammonia is used as adsorbent/adsorbate pair
- Collector type ETSC
- SADS cooling effect (capacity) 25–80 W/kg
- COP is 0.12–0.24
- It is observed that using thermally activated controls to adjust the heating/cooling of the adsorbent bed gave better thermal performance than relying on the solar cycle.

4. Concluding Remark

This book chapter explored solar thermal energy-driven air cooling systems such as absorption, adsorption, and solid desiccant systems. Each type's principal operating technique is explained, as well as some recent advances. Furthermore, a quick comparison is made between them using many inherent indications. Several strategies and parameters are tested for efficacy. Based on capacity and resource availability, this

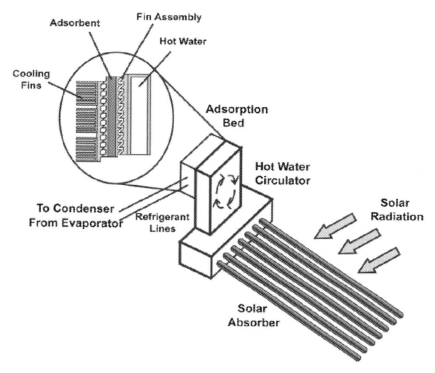

Fig. 13.   System Diagram. *Source:* Reprinted from Robbins and Garimella (2020) Copyright 2020, with permission from Elsevier.

book chapter assists in the selection of a suitable and efficient air-conditioning system. When parameters and working fluids are employed effectively, efficiency and performance are improved. As a result, fuel or energy consumption is reduced, and sustainable growth is promoted. As a result, this chapter also examines the benefits and drawbacks of utilizing a certain parameter.

References

Akhtar, S., Khan, T. S., Tariq, S., Ilyas, S., & Alshehhi, M. S. (2015). Feasibility and basic design of solar integrated absorption refrigeration for an industry. *Energy Procedia, 75,* 508–513. https://doi.org/10.1016/j.egypro.2015.07.441

Al-Yasiri, Q., Szabó, M., & Arıcı, M. (2022). A review on solar-powered cooling and air-conditioning systems for building applications. *Energy Reports, 8,* 2888–2907. https://doi.org/10.1016/j.egyr.2022.01.172

Alahmer, A., Wang, X., & Alam, K. C. A. (2020). Dynamic and economic investigation of a solar thermal-driven two-bed adsorption chiller under Perth climatic conditions. *Energie, 13*. https://doi.org/10.3390/en13041005

Almohammadi, K. M., & Harby, K. (2020). Operational conditions optimisation of a proposed solar-powered adsorption cooling system: Experimental, modeling, and optimisation algorithm techniques. *Energy, 206*. https://doi.org/10.1016/j.energy.2020.118007

Ayadi, O., & Al-Dahidi, S. (2019). Comparison of solar thermal and solar electric space heating and cooling systems for buildings in different climatic regions. *Solar Energy, 188*, 545–560. https://doi.org/10.1016/j.solener.2019.06.033

Bilgili, M. (2011). Hourly simulation and performance of solar electric-vapor compression refrigeration system. *Solar Energy, 85*(11), 2720–2731. https://doi.org/10.1016/j.solener.2011.08.013

Chen, Q. F., Du, S. W., Yuan, Z. X., Sun, T. B., & Li, Y. X. (2018). Experimental study on performance change with time of solar adsorption refrigeration system. *Applied Thermal Engineering, 138*, 386–393. https://doi.org/10.1016/j.applthermaleng.2018.04.061

Dorota, C., Grzebielec, A., & Rusowicz, A. (2014). Solar cooling in buildings. *Technical Transactions Civil Engineering, 3-B*, 65–73.

Faizan, S., Muhammad, S., Takahiko, M., Bidyut, S., Ahmed, A., Imran, A., Yuguang, Z., Riaz, A., & Redmond, S. (2020). Recent updates on the adsorption capacities of adsorbent-adsorbate pairs for heat transformation applications. *Renewable and Sustainable Energy Reviews, 119*. https://doi.org/10.1016/j.rser.2019.109630

Jribi, S., Saha, B. B., Koyama, S., & Bentaher, H. (2014). Modeling and simulation of an activated carbon–CO_2 four bed based adsorption cooling system. *Energy Conversion and Management, 78*, 985–991. https://doi.org/10.1016/j.enconman.2013.06.061

Lattieff, F. A., Atiya, M. A., & Al-Hemiri, A. A. (2019). Test of solar adsorption air-conditioning powered by evacuated tube collectors under the climatic conditions of Iraq. *Renewable Energy*. https://doi.org/10.1016/j.renene.2019.03.014

Lisa, C., & Stephen, H. (2012). Performance evaluation of a liquid desiccant solar air conditioning system. *Energy Procedia, 30*, 542–550. https://doi.org/10.1016/j.egypro.2012.11.064

Li, Z., & Sumathy, Z. (2000). Technology development in the solar absorption air-conditioning systems. *Renewable and Sustainable Energy Reviews, 4*, 267–293. https://doi.org/10.1016/S1364-0321(99)00016-7

Liu, Y. M., Yuan, Z. X., Wen, X., & Du, C. X. (2021). Evaluation on performance of solar adsorption cooling of silica gel and SAPO-34 zeolite. *Applied Thermal Engineering, 182*. https://doi.org/10.1016/j.applthermaleng.2020.116019

Ramli, M. A., Jerai, F., Remeli, M. F., & Yahaya, N. A. (2019). Evaluation of evaporator performance for solar adsorption cooling system. *International Journal of Advanced Research in Engineering Innovation, 1*(2), 42–50. http://myjms.moe.gov.my/index.php/ijaref

Robbins, T., & Garimella, S. (2020). An autonomous solar driven adsorption cooling system. *Solar Energy, 211*, 1318–1324. https://doi.org/10.1016/j.solener.2020.10.068

Roumpedakis, T. C., Vasta, S., Sapienza, A., Kallis, G., Karellas, S., Wittstadt, U., Tanne, M., Harborth, N., & Sonnenfeld, U. (2020). Performance results of a solar adsorption cooling and heating unit. *Energies*, *13*. https://doi.org/10.3390/en13071630

Saidur, R. (2009). Energy consumption, energy savings, and emission analysis in Malaysian office buildings. *Energy Policy*, *37*(10), 4104–4113. https://doi.org/10.1016/j.enpol.2009.04.052

Sha, A. A., & Baiju, V. (2020). Thermodynamic analysis and performance evaluation of activated carbon-ethanol two-bed solar adsorption cooling system. *International Journal of Refrigeration*. https://doi.org/10.1016/j.ijrefrig.2020.12.006

Tso, C. Y., Chan, K. C., Chao, C. Y., & Wu, C. L. (2015). Experimental performance analysis on an adsorption cooling system using zeolite 13X/CaCl$_2$ adsorbent with various operation sequences. *International Journal of Heat and Mass Transfer*, *85*, 343–355. https://doi.org/10.1016/j.ijheatmasstransfer.2015.02.005

Wajid, N. M., Abidin, A. Z., Hakemzadeh, M., Jarimi, H., Fazlizan, A., Fauzan, M. F., Ibrahim, A., Ali, H. A., Waeli, A., & Sopian, K. (2021). Solar adsorption air conditioning system – Recent advances and its potential for cooling an office building in tropical climate. *Case Studies in Thermal Engineering*, *27*. https://doi.org/10.1016/j.csite.2021.101275

Zhai, X. Q., & Wang, R. Z. (2009). A review for absorbtion and adsorbtion solar cooling systems in China. *Renewable and Sustainable Energy Reviews*, *13*(6), 1523–1531. https://doi.org/10.1016/j.rser.2008.09.022

Zhu, L. Q., Tso, C. Y., Chan, K. C., Wu, C. L., Chao, C. Y. H., Chen, J., He, W., & Luo, S. W. (2018). Experimental investigation on composite adsorbent – Water pair for a solar-powered adsorption cooling system. *Applied Thermal Engineering*, *131*, 649–659. https://doi.org/10.1016/j.applthermaleng.2017.12.053

Chapter 4

Evaluating Water-Based Trombe Walls as a Source of Heated Water for Building Applications

Harmeet Singh, Fatemeh Massah and Paul G. O'Brien

Abstract

In this chapter the potential to use water-based Trombe walls to provide heated water for building applications during the summer months is investigated. Design Builder software is used to model a simple single-story building with a south-facing Trombe wall. The effects of using different thermal storage mediums within the Trombe wall on building heating loads during the winter and building cooling loads during the summer are modeled. The amount of thermal energy stored and temperature of water within the thermal storage medium during hot weather conditions were also simulated. On a sunny day on Toronto, Canada, the average temperature of the water in a Trombe wall integrated into a single-story building can reach ~57°C, which is high enough to provide for the main hot water usages in buildings. Furthermore, the amount of water heated is three times greater than that required in an average household in Canada. The results from this work suggest that water-based Trombe walls have great potential to enhance the flexibility and utility of Trombe walls by providing heated water for building applications during summer months, without compromising performance during winter months.

Keywords: Trombe wall; energy storage; building energy; heated water; building envelope; solar energy

1. Introduction

Global warming due to anthropogenic greenhouse gas emissions continues to be a modern-day problem of utmost importance. The Paris Agreement, initiated at the Conference of the Parties in 2021, is a legally binding international treaty on

Pragmatic Engineering and Lifestyle, 61–90
Copyright © 2023 Harmeet Singh, Fatemeh Massah and Paul G. O'Brien
Published under exclusive licence by Emerald Publishing Limited
doi:10.1108/978-1-80262-997-220231004

climate change that brings together 196 parties to commit to limiting global warming to well below 2°C (The Paris Agreement). However, global warming continues to rise due to greenhouse gas emissions, which are strongly tied to the generation and consumption of energy that supports modern-day life.

The building sector is responsible for a significant portion of energy consumption in most countries worldwide (Salari & Javid, 2016). Unfortunately, recent trends indicate that decarbonization in the building sector is not on track to meet the goals set in the Paris Agreement (United Nations Environment Programme, 2021). The significant uses of energy in the residential sector are typically for space and water heating in buildings. As reported by the IEA, out of 28.6 EJ of residential energy consumption, globally, a total of 15.4 EJ (more than 50% of the total energy) is consumed just for space heating purposes (IEA, 2023). The growing energy demands for space and water heating have resulted in 12% of the CO_2 emissions worldwide due to the consumption of fossil fuels (Abergel & Delmastro, 2020).

One approach to reduce greenhouse gas emissions in the building sector is to use passive design strategies such as thermal energy storage, evaporative cooling, radiative cooling, and solar heating (Rodriguez-Ubinas et al., 2014). One promising passive design strategy is to integrate Trombe walls into building facades. Trombe walls store solar thermal energy during the day which can then be transferred to the interior of the building. Trombe walls can provide an effective source of emission-free heating; however, they can also cause overheating during summer months. In this work simulations are performed using Design Builder software to investigate the potential of integrating water in Trombe walls to store thermal energy and to generate heated water. During summer months, when indoor space heating is not required, heated water can be removed from the Trombe wall to prevent overheating and potentially used for building applications such as supplying water for laundry or preheated water to a hot-water tank. In this chapter, background information about Trombe walls is given in Section 2. The methods used to perform the simulations are described in Section 3. The results are presented in Section 4 and discussed in Section 5, while conclusions are provided in Section 6.

2. Background

There are various passive solar technologies available such as solar chimneys, solar roofs, and Trombe walls. Among these technologies, the Trombe wall is a classic passive solar technology that is grabbing attention because it is simple to build, has no operational costs, is highly efficient, and is cost-effective. It has been reported that a Trombe wall can reduce a building's heating demands by 30% (Hu et al., 2017; Long et al., 2018; Saadatian et al., 2012).

There are different types of Trombe walls including classic Trombe walls, composite Trombe walls, Trombe walls with phase change materials (PCMs), water-based Trombe walls, fluidized Trombe walls, electrochromic Trombe walls (Pittaluga, 2013), and Trombe walls with translucent insulation materials

(Baetens et al., 2011; Saadatian et al., 2012; Wang et al., 2020). The classic Trombe wall was first invented and patented by Edward Morse in 1881. However, Felix Trombe and Jacque Michel were the first to promote the Trombe wall by building it in Odeillo, France, in 1967. The classic Trombe wall, shown in Fig. 1a, consists of a glass covering or glazing, an air gap, a thermal storage wall, and may also have upper and lower vents. The thermal storage wall is made of high thermal capacity materials such as bricks, concrete, and stone. The surface of the wall is often painted black to increase the absorption of solar energy. Incident solar energy is transmitted through the glass and generates heat when it is absorbed by the thermal storage wall. A portion of the generated heat increases the temperature of the air within the channel between the outer glass and the thermal storage medium. Due to the density difference between colder air in the interior of the building and hotter air in the air channel, a space heating cycle is created. Heated air moves from the Trombe wall air channel to the interior building space through the top vent and colder air flows from the room into the air channel through the bottom vent as shown in Fig. 1a. The vents are closed when the outside temperature is low or there is no solar energy available, to prevent air flow in the reverse direction. The thermal storage wall absorbs direct and diffused solar energy during the day which is then released gradually by convection or conduction. The thickness of the air gap between glass and thermal storage wall is typically in the range of 1–10 cm.

Over time, several modifications to the design of the classic Trombe wall were proposed to improve performance such as adding shades or blinds to prevent overheating during summer, adding thermal fins to the internal wall to increase heat transfer, and building a zigzag-shaped Trombe wall to increase the heating effect during colder periods and to prevent overheating during the day (Hu et al., 2017; Saadatian et al., 2012; Wang et al., 2020). For example, to overcome the low thermal resistance of the classic Trombe wall, a composite Trombe wall, also known as a Trombe–Michel wall was proposed. As shown in Fig. 1b, the Trombe–Michel wall comprises an outer glass cover, a nonventilated air channel between the glass cover and the thermal storage wall, and a vented air channel between the thermal storage wall and a vented insulation layer. In the Trombe–Michel wall configuration solar energy is absorbed and a greenhouse effect is generated in the outer air channel between the thermal storage wall and glass covering, which then heats the storage wall, which in turn heats the room as air cycles through the inside air channel. Chen et al. modified the composite Trombe wall by adding a porous absorber in between the thermal storage wall and glass covering. The porous absorber is treated as a thermal insulator and heat storage buffer which absorbs the thermal energy first and then passes it to the thermal storage wall (Chen & Liu, 2004, 2008; Wang et al., 2020).

The thermal storage capacity of the classical Trombe wall is limited because it stores energy in the form of sensible heat. Therefore, another strategy to improve the performance of Trombe walls is to integrate PCMs into the thermal storage wall, as shown in Fig. 1c and d, to store latent heat and increase the heat storage capacity. Under ideal operating conditions the PCM melts when it absorbs solar thermal heat during the day and releases latent heat when it solidifies at night.

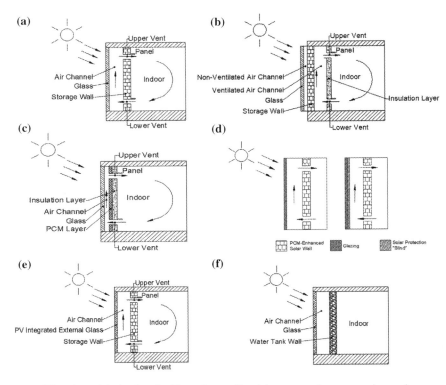

Fig. 1.   (a) A classic Trombe wall with outer glass covering, air channel, thermal storage wall, and vents. (b) A composite Trombe wall with outer glazing, nonventilated air channel, storage wall, ventilated air channel and vents. (c) A PCM Trombe wall with outer glass covering, air channel, PCM layer integrated on the outside surface of insulation wall with vents for latent heat storage. (d) The working mechanism of a PCM Trombe wall. (e) A PV Trombe wall with PV integrated outer glass covering, air channel and thermal energy storage wall with vents. (f) A water Trombe wall with outer glass covering, air channel, and a water tank wall as a storage medium.

Furthermore, the classical Trombe wall can pose architectural challenges because it increases the dead load of a building (Omara & Abuelnuor, 2020), whereas PCMs store more energy in smaller volumes, thereby providing a lightweight alternative to heavy Trombe walls. Bourdeau conducted an experimental study to show that a 15-cm thick concrete wall can be replaced by a 3.5-cm thick PCM wall to deliver a similar performance (Bourdeau, 1980). Also, Gracia et al. performed experiments to analyze the thermal performance of SP 22 micro-encapsulated PCMs, which have a phase change temperature of 20°C. The experimental setup consisted of two rooms of 2.4 × 2.4 × 5.1 m dimensions in Spain. They integrated PCMs in the ventilated air channel in one of the rooms. It was observed in the experiments that a ventilated Trombe wall with PCMs performed better than the one without PCMs. The temperature of the room that did not have a PCM Trombe wall dropped daily during the testing period according to the outer weather conditions, but the temperature of the room with the PCM Trombe wall increased from 9°C to 18°C during the same testing period, thus reducing the heating, ventilation, and air conditioning (HVAC) heating load (De Gracia et al., 2012). Leang et al. compared the performance of two composite walls: a composite Trombe wall consisting of a concrete thermal storage wall and a composite Trombe wall consisting of a commercial PCM, micronal. Both numerical and experimental studies were performed to compare the performance of the two walls. The results showed that the PCM thermal storage wall has more than 50% additional heat storage capacity as compared to the concrete storage wall. Also, a PCM wall of 4-cm thickness was more than four-times effective in charging and distributing the warmth to the room as compared to a 15-cm thick concrete wall (Leang et al., 2017). However, there are some challenges in using PCMs in Trombe walls. The optimal quantity of PCM required in a Trombe wall is challenging to determine. A low amount of PCM can decrease the performance whereas a high amount can cause overheating due to increased thermal load. The optimal phase change temperature of a PCM is also a challenge to determine because a high phase change temperature can cause overheating and a low phase change temperature can reduce the inside temperature. Furthermore, materials used to encapsulate PCMs are sometimes not compatible with medium temperature PCM systems. Also, PCMs may corrode and have low mechanical strength at high temperatures. Therefore, further research is required to improve the durability of PCMs or to prohibit these conditions (Omara & Abuelnuor, 2020).

Another strategy is to integrate photovoltaic (PV) cells into Trombe walls to generate electric power and increase their overall utility. A PV Trombe wall may consist of PV modules integrated onto the external glass cover as shown in Fig. 1e. The heat gain in the PV Trombe wall is lower than that of the classical Trombe wall because the PV panel obstructs the transmission of some of the solar radiation into the air channel. However, the benefits of producing electric power using the PV cells often outweighs the disadvantages of reducing the amount of solar energy reaching the storage wall, especially considering that semitransparent PV cells can be used to minimize these loses (Ozin et al., 2011; Yang et al., 2013). Furthermore, PV Trombe walls can also prevent or reduce overheating during the summer season (Saadatian et al., 2012; Wang et al., 2020).

Another highly promising approach to improving the performance of Trombe walls is to use water as part of the thermal storage wall, as shown in Fig. 1f. Water-based Trombe walls offer a number of advantages. Firstly, water has a high heat capacity and can store more thermal energy than other Trombe walls on a per volume basis. The volumetric heat capacity of water is 4,186 kJ/(m^3·K), whereas the volumetric heat capacity of concrete is about 2,000 kJ/(m^3·K) and sand is about 2,600 kJ/(m^3·K). The high heat capacity also keeps the temperature of the water Trombe wall lower than other Trombe walls, which can help reduce thermal losses to the surroundings. Also, the water can potentially be removed from the Trombe wall and used as a source of hot water for the building. The heat energy absorbed by the water is distributed by convection and hot water can be removed from the top of the tank and supplied to the building for various purposes. The ability to remove hot water from the Trombe wall during summer months is especially attractive because many Trombe walls cause buildings to overheat during the summer season. Furthermore, the transparent property of water can be utilized in making esthetically pleasing designs when architecting buildings. Indeed, translucent materials of different colors have been integrated into water Trombe walls, giving them a vibrant appearance (Solarcomponents, 2019).

Despite their unique combination of advantages, a limited amount of research has been done on water Trombe walls in comparison to other Trombe walls. Weiliang et al. analyzed the thermal performance of a Trombe wall consisting of water as a thermal energy storage medium. A south-facing water Trombe wall prototype with a single-story house with floor area of 700 m² and shape coefficient 0.374 (ratio of external surface area and inner volume) was set up in North China for experimental analysis, and a simulation model was set up in TRNSYS. The outer layer of the Trombe wall was made of a steel plate. A total of 29 small modules of water (with dimensions of 1.1 m × 0.4 m × 2.5 m) were placed along the inner side of the wall. Numerical and experimental analysis were conducted and compared with the traditional Trombe wall. The results showed a reduction in energy consumption per year by 8.6% and the indoor thermal comfort evaluation index was improved by approximately 13% as compared to the classic Trombe wall (Wang et al., 2013).

Nayak did a numerical analysis of thermal performance between a south-facing drum water wall and a water transwall. The drum wall consisted of metallic containers of water stacked upon each other. One surface of the wall is colored black and has glazing on it, but the other surface can be separated from the room by a concrete wall or insulating layer. The Transwall consists of water in containers that are made of parallel glass walls. Each wall has a semitransparent material kept either between the water column and the room, between glazing and the water column, or in the water column itself (Transwall). Energy balance equations were used to solve the Fourier equations of heat conduction for both configurations to calculate the heat flux entering the room. The results show that the Transwall meets the daytime heating load more effectively than a drum wall whereas a drum wall performs better in terms of load leveling and day–night performance (Nayak, 1987).

Adams et al. conducted a study to investigate the optimal thickness of a water storage wall. A controlled environment was built in the lab to analyze the effects of a heat source on the water wall with 3″, 6″, and 9″ thicknesses. The environment consisted of three parts: (1) exterior space, (2) interior space, and (3) the water wall. The dimensions of exterior and interior volumes were 1 ft³ (1 ft × 1 ft × 1 ft). The exterior volume consisted of a heat source (halogen bulb), and the interior volume was a vacant space. The heat source was turned on for 5 hours and then switched off to monitor the thermal performance of exterior and interior volumes over time. It was observed that 6″ and 9″ thick walls performed better than 3″ wall.

The thinner wall allowed the indoor temperature to increase at a faster rate and maintained high temperature for longer periods of time than the thick walls. Thus, 3″ wall was not able to efficiently regulate the room temperature than 6″ and 9″ walls. Also, the heat transfer through 3″ wall was higher than that of 6″ and 9″ walls. It was concluded that the Trombe wall thickness should at least be greater than 3″ for optimal performance (Adams et al., 2010).

Kaushik and Kaul proposed a thermal storage mixed water-mass wall and performed a heat transfer analysis. The analysis was performed on a room on the ground floor in a building without any air conditioning. The room consisted of a south-facing wall made of concrete and water layers (concrete–water–concrete) for heat gain and space heating purposes into the room. The periodic heat transfer analysis included the heat transfer through the walls and roof, heat conduction to the floor and furnishings, and heat loss due to infiltration and ventilation. The outer walls and roof of the room were exposed to sun and atmospheric air temperature, and inner walls were in contact with changing air temperature. Numerical analysis was performed to study the changes in indoor air temperature and heat fluxes based on hourly solar irradiance, and atmospheric temperature for a mild day of winter in New Delhi, India. They considered different configurations of storage walls consisting of concrete–water–insulation. The results showed that the concrete–water–concrete wall performed better than other mixed storage walls such as a water wall with insulation only for space heating and maintaining a comfortable temperature. The placement and thickness of concrete or insulation have considerable effect on the heat flux and inside air temperature. Varying the thickness of the insulation rather than the thickness or placement of the concrete walls generated the desirable results (Kaushik & Kaul, 1989).

Turner et al. studied the performance of a water Trombe wall consisting of a 7.6-cm diameter plastic tube embedded into a studded wall. During winter season, the system was charged for six hours with hot ambient air and then passively discharged for 18 hours. The results showed that the Trombe wall temperature remained approximately 2.6°C higher even after the discharging cycle completed which helped in reducing the heating load of the house. Also, it was noted that during summer season, water walls can be charged using cool ambient air at night to achieve thermal comfort during daytime (Turner et al., 1994).

Tiwari et al. did a comparative study of total heat gain of different south-facing Trombe walls such as a glass wall, water wall, active air collector wall, and a transwall. Several design parameters were taken into consideration such as the thickness of the water wall and transwall, and the flow rate of the air collector wall. The design parameters were varied for a heated room under winter weather conditions to conduct different experiments. The results showed that the performance of the water wall and the transwall is better for space heating during night time because of their greater thermal storage capacity. On the other hand, the glass wall and the air collector wall are efficient for space heating during sunshine hours. It was also observed that the air temperature of the room was higher during the extreme winter conditions of Srinagar, India, when a transwall was used in a nonconditioned passive solar house in comparison to when a water wall is used (Upadhya et al., 1991).

Mohamad et al. designed a novel Trombe wall to reduce the heating and cooling loads in the winter and summer seasons, respectively. The Trombe wall consists of a water tank that acts as a thermal energy storage medium and can also supply hot water if required. The proposed Trombe wall model can be used for space heating during day and night, and excess heat can be used for domestic hot water supply in the summer season. The excess heat can be extracted out of the building through vents or can be used for domestic water heating purposes to reduce the cooling load. The experimental results showed that the proposed system is more efficient in charging and discharging thermal energy in comparison to a classical Trombe wall. The thermal storage efficiency of the proposed system is more than 70%. Furthermore, a numerical analysis was performed to study the effect of the heat transfer coefficient. The study showed that the air-side heat transfer coefficient is also an important factor that needs to be depicted correctly in the model. It can be increased by adding additional extended surfaces such as (fins) or turbulent generators, but it will add weight and cost to the model. It was also observed that decreasing the water tank size increases the temperature but the amount of energy stored does not change (Mohamad et al., 2019).

In the aforementioned work done by Mohamad et al. (2019), an opaque solar absorber was used at the outer surface of the water storage wall. Recently, Singh et al. used a lab-scale Trombe wall prototype to demonstrate that tinted acrylic sheets can be used as the absorber in semitransparent water-based Trombe walls (Singh & O'Brien, 2022). Herein, the potential of using a semitransparent absorber in a water storage wall is numerically investigated. Specifically, a tinted acrylic sheet is integrated into the water wall, such that it retains its transparency to some degree, which is desirable for the design of esthetically pleasing Trombe walls. In this work, Design Builder software is used to simulate the heating and cooling loads on a simple single-story building with a Trombe wall. The effects of using water-based thermal storage configurations within the Trombe wall are evaluated. The results from the simulation show that the water-based Trombe walls perform similarly to Trombe walls during the winter season. Furthermore, during the summer season large amounts of solar thermal energy can be stored in water-based Trombe walls, and the average temperature of the water can be elevated to more than 55°C. The temperature of the stored heated water is high enough to be used for numerous building applications including faucets, laundry, and preheated water for the buildings hot water tank.

3. Methods

3.1 Description of the Building Simulated in Design Builder Software

As shown in Fig. 2, a three-dimensional (3D) model of a single-story building with a Trombe wall is created in Design Builder simulation software to numerically investigate the performance of different Trombe wall configurations.

As shown in Fig. 3a, the width, depth, and height of the building are 4,000 mm, 3,360 mm, and 3,000 mm, respectively. The model consists of two zones: a

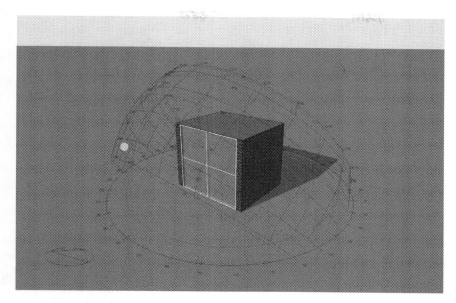

Fig. 2.   Three-Dimensional Model of a Single-Story Building With a
Trombe Wall.

room (occupied zone) and a Trombe wall with an air channel (Trombe wall zone).
The occupied zone room is set to the standard zone type and the Trombe wall
zone is set to the cavity zone type and no occupancy is considered in Design
Builder. The air changes per hour (ACH) due to ventilation is set as 5 for the
occupied room zone and 0 for the Trombe wall zone. In practice, the recommended ACH depends on the type of building being considered and the number
of occupants, and ranges from 0.35 for a multifamily unit to 3.5 for classrooms
and as high as 20 for operating rooms. However, it has recently been recommended that minimum ACH levels should be set from 4 to 6 ACH to reduce
airborne transmission of infectious diseases (Allen & Ibrahim, 2021). Thus, in this
work we assume the ACH of the occupied zone is 5 in anticipation of increasing
ACH values in the future and also to investigate building energy loads under
more demanding conditions wherein the ACH is relatively high compared to
current standards.

   The occupied zone and the air channel zone are separated by a partition wall
that functions as a thermal energy storage medium (Fig. 3b). The Trombe wall
zone consists of a sun-facing high transmissivity glazing, an air gap, and a
partition wall consisting of a thermal energy storage medium and two vents
(upper and lower as shown in Fig. 3c) to facilitate airflow in and out of an air
channel located between the partition wall and the glazed window. Incident solar
energy transmitted through the outer glazing is absorbed at the surface of the

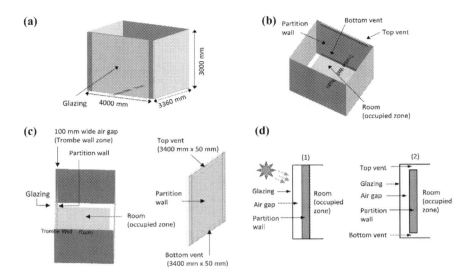

Fig. 3.   (a) Model of the single-story building showing the glazing in front of the Trombe wall. (b) Model of the room (the occupied zone) within the single-story building showing the inside of the partition wall and the top and bottom vents. (c) The 100 mm wide air gap (which is the Trombe wall zone) between the partition wall and the glazing for the building shown in (a). (d) Schematic diagram of the side view of the partition wall which contains the thermal energy storage medium for the (1) summer and (2) winter cases.

partition wall. Solar thermal energy generated in the partition wall heats the thermal energy storage medium and the air within the air channel (Fig. 3d). Warm air can be delivered to the occupied zone through the upper vent. Heat can also be transferred directly from the inside surface of the thermal storage wall to the room and thus stored thermal energy can be used at a later time when sunshine is unavailable. Moreover, when water is used as the thermal energy storage medium, heated water can be removed and used for other purposes within the building, such as in appliances, hydronic radiant heating systems at other locations in the building, and for domestic hot water supply. To date, little research has been done in this area.

3.2 Properties of Materials Selected in the Model

The effects of altering the thicknesses and materials used for the thermal energy storage wall (the partition wall shown in Fig. 3b–d) on the building cooling and heating loads, and thermal energy stored, are compared. The details about simulating the building shown in Fig. 2 in Design Builder are provided below.

3.2.1 Trombe Wall

The Trombe wall has a double-layered glazing comprising 6-mm thick generic clear glass separated by a 13 mm thick air gap. There is a 100 mm thick air gap between the glazing and the partition wall. For this glazing, the solar heat gain coefficient (SHGC) is 0.703, the direct solar transmission is 0.604, the light transmission is 0.781, and the U-value is 2.665 W/m²·K. The partition wall is used as the thermal energy storage wall in the simulations (the thermal storage wall is described in Section 3.2.5). The upper and lower vents in the partition wall have a width of 3,400 mm and a height of 50 mm. Both vents are set to be "ON" during the winter conditions. The timing and duration of vent openings is controlled based on the room and Trombe zone temperatures. The vents are opened when the room temperature is below the Trombe zone temperature. Both vents are always closed during summer conditions.

3.2.2 Walls

The walls of the single-story building shown in Fig. 2 are constructed with four layers, and the thickness, thermal properties, and radiative properties used in the simulations are shown in Table 1. A_T, A_S, and A_V are the thermal absorptance (emissivity), solar absorptance, and the absorptance for visible light, respectively. Also, λ is the thermal conductivity, C_p is the specific heat capacity, and ρ is the density.

3.2.3 Roof

The roof of the model building shown in Fig. 2 has four layers, and the thickness, thermal properties, and radiative properties of each layer are shown in Table 2.

3.2.4 Floor

3.2.4.1 Ground Floor. The ground floor of the building shown in Fig. 2 is made of four layers with the thickness, thermal properties, and surface properties shown in Table 3 (Team DesignBuilder, 2019).

3.2.5 Thermal Energy Storage Medium

Simulations are performed to analyze the performance of Trombe walls with different thermal energy storage mediums during summer and winter conditions (See Section 3.3). The different cases of Trombe walls simulated for winter and summer conditions are described in Tables 5 and 6, respectively. Simulations are carried out in winter weather conditions to compare the performance of water-based Trombe walls to that of Trombe walls made from more conventional materials. In some of these configurations, water is chosen as a thermal energy storage medium and sand is used in other configurations for comparison. In other configurations, both sand and water are used as inner and outer surfaces in the

Table 1. Properties of Construction Material Used for the Walls in the Building Shown in Fig. 2.

Construction Material	Thickness	Thermal Properties	Radiative Properties
4 layers			
Outermost layer: Bricks	100 mm	λ (W/m·K): 0.84	A_T: 0.9
		C_p (J/kg·K): 800	A_S: 0.7
		ρ (kg/m³): 1700	A_V: 0.7
Layer 2: XPS extruded polystyrene	79.40 mm	λ (W/m·K): 2.8	A_T: 0.9
		C_p (J/kg·K): 1,000	A_S: 0.6
		ρ (kg/m³): 2,600	A_V: 0.6
Layer 3: Concrete block	100 mm	λ (W/mK): 2.8	A_T: 0.9
		C_p (J/kg·K): 1,000	A_S: 0.6
		ρ (kg/m³): 2,600	A_V: 0.6
Innermost layer: Gypsum plastering	13 mm	λ (W/m·K): 2.8	A_T: 0.9
		C_p (J/kg·K): 1,000	A_S: 0.5
		ρ (kg/m³): 2,600	A_V: 0.5

same thermal energy storage medium to investigate the effects of using multi-layered thermal storage mediums.

The simulations performed under summer weather conditions are used to evaluate the potential benefits of using water-based Trombe walls to generate hot water during summer months which can then be removed from the Trombe wall and used for building applications. The materials used in these Trombe wall configurations are water, tinted acrylic sheets, and transparent insulation. Tinted acrylic is used to absorb a portion of the incident sunlight to generate heat that can be stored in the water while transmitting a portion of the visible sunlight such that the Trombe wall can have a semitransparent appearance. That is, the water-based Trombe walls investigated herein can be used in the design of tinted "see-through" walls, as it is important for Trombe walls to be esthetically pleasing to promote their widespread integration into buildings. The insulation used on some thermal storage mediums for the summer weather conditions is assumed to be aerogel insulation which has high

Table 2. Properties of Construction Material Used for the Roof of the Model Building Shown in Fig. 2.

Construction Material	Thickness	Thermal Properties	Radiative Properties
Outermost layer: Asphalt 1	10 mm	λ (W/m·K): 0.7	A_T: 0.9
		C_p (J/kg·K): 1,000	A_S: 0.85
		ρ (kg/m³): 2,100	A_V: 0.9
Layer 2: Medium weight (Mw) glass wool (rolls) polystyrene	100 mm	λ (W/m·K): 0.04	A_T: 0.9
		C_p (J/kg·K): 840	A_S: 0.6
		ρ (kg/m³): 12	A_V: 0.6
Layer 3: Air gap	200 mm	R (m²·K/W): 0.1800	A_T: 0.9
			A_S: 0.6
			A_V: 0.6
Innermost layer: Plasterboard	13 mm	λ (W/m·K): 0.25	A_T: 0.9
		C_p (J/kg·K): 896	A_S: 0.5
		ρ (kg/m³): 2,800	A_V: 0.5

transmittance and low thermal conductivity. It allows for 90% of the solar radiation to pass through and can have a U-value of 0.1 W/m²·K (Baetens et al., 2011; Schultz & Jensen, 2008). The properties of the sand, water, tinted acrylic and aerogel insulation used in the simulations are provided in Table 4. The amount of thermal energy stored in the layers within the thermal storage mediums is determined using the heat capacity and the increase in the average temperature of the layer under consideration between sunrise and 2 p.m. The average temperature of a give layer is assumed to be the average temperature of its surfaces. Also, a limitation to the simulations presented in this chapter is that the water is treated as a solid rather than a fluid. That is, hotter water does not circulate toward the top of the water storage medium. The implications of this assumption are that the results from the simulations underestimate the temperature of the water at the top of the water storage medium.

Table 3. Properties of Construction Materials Used for the Ground Floor of the Model Building Shown in Fig. 2.

Construction Object	Construction Material	Thickness	Thermal Properties	Surface Properties
Ground floor (composed of four layers)	*Outermost layer:* Urea-formaldehyde foam	100 mm	λ (W/m·K): 0.04 C_p (J/kg·K): 1,400 ρ (kg/m³): 10	A_T: 0.9 A_S: 0.6 A_V: 0.6
	Layer 2: Cast concrete	100 mm	λ (W/m·K): 1.13 C_p (J/kg·K): 1,000 ρ (kg/m³): 2000C	A_T: 0.9 A_S: 0.6 A_V: 0.6
	Layer 3: Floor/Roof screed	70 mm	λ (W/m·K): 0.41 C_p (J/kg·K): 840 ρ (kg/m³): 1200C	A_T: 0.9 A_S: 0.73 A_V: 0.73
	Innermost layer: Timber flooring	30 mm	λ (W/m·K): 0.14 C_p (J/kg·K): 1,200 ρ (kg/m³): 650C	A_T: 0.9 A_S: 0.78 A_V: 0.78

Table 4. Thermal Properties of the Sand, Water, Tinted Acrylic, and Aerogel Insulation.

Properties / Materials	λ (W/m·K)	C (J/kg·K)	ρ (kg/m³)
Water	0.63	4,190	990
Tinted acrylic	0.8	800	2,500
Sand	2.8	1,000	2,600
Aerogel	0.014	1,000	150

3.3 Weather Conditions Used to Perform the Simulations

The Trombe wall faces south and hourly weather files (WYEC2) for Toronto, Ontario, are used to conduct the simulations. The simulations are performed for a duration of two months under summer weather conditions (from July 1st to August 31st, 2020) and under winter weather conditions (from January 1st to February 28th, 2021). The results for three consecutive days within these time-frames for the summer and winter weather conditions were selected and presented in this work. The July 14–16, 2020 period is selected for summer simulations and the February 2–4, 2021 period is selected for winter simulations. The simulation periods were selected after comparing the average solar irradiance of these three days over the summer and winter months. The three-day average over July 14, 15, and 16 is 877 W/m² and is the highest over the months of July and August. The average solar irradiance over the three days of February 2, 3, and 4 is 843 W/m² and this is the highest over the months of January and February. A thermostat is used to control the temperature in the room zone. For simulations carried out during summer weather conditions, the cooling setpoint and setback temperatures are 20°C and 22°C, respectively. For simulations carried out under winter weather conditions, the heating setpoint and setback temperatures are 24°C and 20°C, respectively. Notably, the cooling setpoint of 20°C is on the low-end of the range of setpoints typically used in buildings during the summer. Also, the heating setpoint of 24°C is on the high-end of the range of setpoints normally used during the winter. Selecting these values results in relatively high cooling and heating demands during the summer and winter, respectively. This is beneficial for simulating the performance of the Trombe wall under demanding conditions (when the occupants would like cooler air temperatures in the summer and higher temperatures in the winter).

3.4 Model Configuration in Energy Plus Software

The simulations are carried out using Design Builder software, which is an advanced user interface for EnergyPlus (Energy Plus). EnergyPlus is the United

States Department of Energy's open-source whole-building modeling engine and can be used to model energy consumption for heating, cooling, ventilation, lighting, and water usage.

3.4.1 Model Options

Design Builder loads default data at the site and building level when a building is created which is known as model data. Model data settings can be overwritten and there are various options or configuration settings available to customize the model at the building level. A summary of the options selected to simulate the building structure shown in Fig. 2 are provided in Table 7 (Team DesignBuilder, 2019).

4. Results

4.1 Comparison of Heating Load During Winter for Different Trombe Wall Configurations

In this work we investigate the potential to enhance the performance and utility of Trombe walls during the summer season by using water as a thermal storage medium. However, we also model the performance of water-based Trombe walls during winter conditions to evaluate their year-round performance. One method of evaluating the performance of Trombe walls during the winter is to monitor their ability to minimize the building heating load. To analyze the performance of the thermal energy storage walls described in Table 5, the total heating load for the building shown in Fig. 2 over three days (Feb 2, Feb 3, and Feb 4, 2021) was simulated for different thicknesses of the thermal energy storage medium. The lowest heating loads achieved for each type of thermal storage medium considered are shown in Fig. 4 (the thicknesses of the material layers that provided the best results for each case are provided in brackets beneath each bar in Fig. 4).

For comparison, the heating load of the building is calculated when there is no Trombe wall (in this case, denoted as Case 0, the Trombe wall at the south side of the building is replaced with an external wall with properties similar to the other external walls as described in Section 3.2.2). As shown by the first bar in Fig. 4, the total heating load is approximately 14.41 MJ/m² when the Trombe wall at the southern side of the building is replaced with an external wall. Of the cases considered in Fig. 4, the building has the highest heating load of about 14.86 MJ/m² when the thermal storage medium is a 40-cm thick water wall. This is because the water wall does not absorb incident solar radiation and store it as thermal energy unless tinted acrylic is inserted within it, such as for Case 3 which has a much lower heating load of 6.64 MJ/m². For all other cases considered, the heating load is less than half that for the case when the pure water Trombe wall is used. A minimal heating load of 6.63 MJ/m² occurs for the case when the thermal storage medium is composed of a 6-cm thick water wall on the outer side and a 17.5-cm thick sand wall on the inner side (e.g., wat(6 cm)/sand(17.5 cm)). Nevertheless, the results shown in Fig. 4 support the concept of using a

Table 5. Properties of the Thermal Energy Storage Wall (Partition Wall) Investigated for Winter Weather Conditions.

Case 1: Water

Water

Material: Water

Surface properties: $A_T = 0.9$, $A_S = 0$, $A_V = 0$

Description: Naming convention: "wat". For example, a 20-cm thick storage wall made of water is denoted as wat(20 cm). In the simulations the thicknesses of the Trombe wall is varied from 5 cm to 40 cm in increments of 5 cm.

Case 2: Water with tinted acrylic at its outer side

Tinted acrylic   Water

Materials: Tinted acrylic and water

Surface properties: Water*: $A_T = 0.9$, $A_S = 0.8$, $A_V = 0.8$
Tinted acrylic: $A_T = 0.9$, $A_S = 0.8$, $A_V = 0.8$

Description: Naming convention: "ta/wat". For example, a storage wall made of 1 cm thick tinted acrylic and a 10-cm thick water wall is denoted as ta(1 cm)/wat(10 cm). In the simulations the thickness of the water is varied from 5 cm to 40 cm with increments of 5 cm. The thickness of the tinted acrylic is 1 cm. The thickness of the insulation layer is 5 cm.

Case 3: Water with tinted acrylic at its center

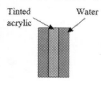

Tinted acrylic   Water

Materials: Water and tinted acrylic

Surface properties: Water*: $A_T = 0.9$, $A_S = 0.8$, $A_V = 0.8$
Tinted acrylic: $A_T = 0.9$, $A_S = 0.8$, $A_V = 0.8$

Description: Naming convention: "wat/ta/wat". For example, the storage wall made of 1 cm thick tinted acrylic and 10 cm thick water wall is denoted as wat(5 cm)/ta(1 cm)/wat(5 cm). In the simulations the thicknesses of the water wall is varied from 5 cm to 20 cm with increments of 5 cm. The thickness of the tinted acrylic is 1 cm.

Case 4: Sand

Sand

Materials: Granite sand

Surface properties:* Granite sand: $A_T = 0.9$, $A_S = 1$, $A_V = 1$

Description: Naming convention: "sand". For example, a 20-cm thick storage wall made of sand is denoted as sand (20 cm). The thicknesses of the simulated Trombe wall varies from 5 cm to 40 cm with increments of 5 cm.

Table 5. *(Continued)*

Case 5: Water at inner and sand at outer sides

Material: Water and granite sand

Surface properties: Water*: $A_T = 0.9$, $A_S = 1$, $A_V = 1$
Granite sand: $A_T = 0.9$, $A_S = 1$, $A_V = 1$

Description: Naming convention: "wat/sand". For example, a 5-cm thick storage wall made of water and a 17.5-cm thick sand wall is denoted as wat(5 cm)/sand(17.5 cm). The total thicknesses of the simulated Trombe walls vary from 20.5 cm to 32.5 cm

Case 6: Sand at inner and water at outer sides

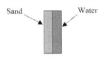

Material: Granite sand and water

Surface properties: Granite sand $A_T = 0.9$, $A_S = 1$, $A_V = 1$
Water: $A_T = 0.9$, $A_S = 1$, $A_V = 1$

Description: Naming convention: "sand/wat". For example, a 17.5-cm thick storage wall made of sand and a 5-cm thick water wall is denoted as sand (17.5 cm)/wat(5 cm). The total thicknesses of the simulated Trombe walls vary from 20.5 cm to 32.5 cm

---

Note: *The properties of the first layer in the partition wall set the absorption for the thermal storage wall in the simulations. When water is the first layer, its absorptive properties are set to that of the absorbing layer within the wall.

water-based Trombe walls for efficient solar energy collection during winter, as a minimal heating load of 6.64 MJ/m² is achieved when the thermal storage medium in the Trombe wall is a tinted acrylic sheet inserted into a water (which is almost identical to the heating load for the wat(6 cm)/sand(17.5 cm) case).

4.2 Comparison of Cooling Load during Summer for Different Water-Based Trombe Wall Configurations

To compare the water-based thermal storage mediums shown in Table 6, the total thickness of the water in these structures is set to 20 cm in the simulations. A thickness of 20 cm is selected since it provides a good balance between having a large amount of thermal energy storage capacity and a reasonably sized wall. Fig. 5 shows the total cooling load for the building shown in Fig. 2 when the Trombe wall consists of a 20-cm thick water storage wall over three consecutive days (July 14, 15, and 16, 2020) in summer for the cases shown in Fig. 5. The cooling load for the building is also calculated when there is no Trombe wall for comparison (for the case when the Trombe wall is absent, denoted as Case 0, the wall at the

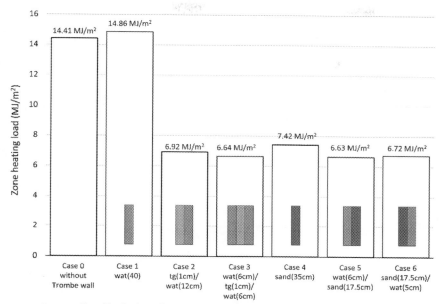

Cases considered for the thermal energy storage medium within the Trombe wall during winter conditions

Fig. 4. Comparison Between the Best Cases of Total Heat Load Normalized by Floor Area (MJ/m²) for Three Consecutive Days (Feb 2, Feb 3, and Feb 4, 2021) in the Room Based on the Optimal Trombe Wall Thickness During Winter in Toronto, Ontario.

south side of the building is an external wall with properties similar to the other external walls as described in Section 3.2.2).

The cooling load is 2.23 MJ/m² for the case when there is no Trombe wall. The cooling load is increased to 9.24 MJ/m² when the thermal storage medium is composed of a 20-cm thick water layer and a 1-cm thick acrylic sheet, located at either the front or the middle of the water container. On the other hand, the cooling load is 2.25 MJ/m² when the storage wall has a 20-cm thick water layer with 5-cm thick insulation layers on either side. When insulation is integrated on both sides of the Trombe wall the cooling load is 2.49 MJ/m² when a tinted sheet resides at the front of the water container. Notably, for the case when the water wall is surrounded by insulation, there is an increase of about 10% in the cooling load when a tinted acrylic sheet is placed at the front side of the water as compared to the case when there is no Trombe wall (2.49 MJ/m² vs. 2.25 MJ/m²). This suggests cooling loads may increases slightly when water-based Trombe walls with tinted acrylic sheets are integrated into the building façade. However, this increase in the cooling load is small in comparison to the thermal energy that

Table 6. Properties of the Thermal Energy Storage Wall (Partition Wall) Investigated for Summer Weather Conditions.

Case 1: Water

Material: Water

Surface properties: $A_T = 0.9$, $A_S = 0$, $A_V = 0$

Description: Naming convention: "wat". For example, a 20-cm thick storage wall made of water is denoted as wat(20 cm). The thicknesses of the simulated Trombe walls vary from 5 cm to 40 cm with increments of 5 cm.

Case 2: Water with transparent insulation at its inside surface

Material: Water and aerogel

Surface properties: Water*: $A_T = 0.9$, $A_S = 0$, $A_V = 0$

Aerogel: $A_T = 0.9$, $A_S = 0$, $A_V = 0$

Description: Naming convention: "wat/ins". For example, a storage wall made of a 5-cm thick water wall and a 5-cm thick insulation wall is denoted as wat(5 cm)/ins(5 cm). The thicknesses of the water layer vary from 5 cm to 40 cm with increments of 5 cm. The thickness of the insulation wall is 5 cm.

Case 3: Water with transparent insulation on both sides

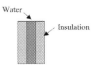

Material: Water and aerogel

Surface properties: Water: $A_T = 0.9$, $A_S = 0$, $A_V = 0$

Aerogel: $A_T = 0.9$, $A_S = 0$, $A_V = 0$

Description: Naming convention: "Ins/wat/ins". For example, a storage wall made of a 5-cm thick water wall with 5-cm thick layers on insulation on both of its sides is denoted as ins(5 cm)/wat(5 cm)/ins(5 cm). The thicknesses of the simulated water walls vary from 5 cm to 40 cm with increments of 5 cm. The thickness of the insulation layers is 5 cm for both the inner and outer sides.

Case 4: Water with tinted acrylic at its outer sides

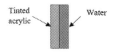

Material: Water and tinted acrylic

Surface properties: Water*: $A_T = 0.9$, $A_S = 0.8$, $A_V = 0.8$

Tinted acrylic: $A_T = 0.9$, $A_S = 0.8$, $A_V = 0.8$

Aerogel: $A_T = 0.9$, $A_S = 0$, $A_V = 0$

Table 6. *(Continued)*

Description: Naming convention: "ta/wat". For example, a storage wall made of 1-cm thick tinted acrylic and 10-cm thick water wall is denoted as ta(1 cm)/wat(10 cm). The thicknesses of the simulated water walls vary from 10 cm to 40 cm with increments of 10 cm. The thickness of the tinted acrylic is 1 cm.

Case 5: Water with tinted acrylic at its outer side and transparent insulation at its inner side

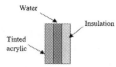

Material: Water, tinted acrylic, and aerogels

Surface properties: Water*: $A_T = 0.9$, $A_S = 0.8$, $A_V = 0.8$

Tinted acrylic: $A_T = 0.9$, $A_S = 0.8$, $A_V = 0.8$

Aerogel: $A_T = 0.9$, $A_S = 0$, $A_V = 0$

Description: Naming convention: "ta/wat/ins". For example, a storage wall made of 1-cm thick tinted acrylic, a 10-cm thick water wall, and a 5-cm thick insulation layer is denoted as ta(1 cm)/wat(10 cm)/ins(5 cm). The thicknesses of the simulated water walls vary from 10 cm to 40 cm with increments of 10 cm. The thickness of the tinted acrylic is 1 cm and the thickness of the insulation is 5 cm for all cases.

Case 6: Water with tinted acrylic at its outer side and transparent insulation on both sides

Material: Water, tinted acrylic and aerogel

Surface properties: Water*: $A_T = 0.9$, $A_S = 0.8$, $A_V = 0.8$

Tinted acrylic $A_T = 0.9$, $A_S = 0.8$, $A_V = 0.8$

Aerogel: $A_T = 0.9$, $A_S = 0.8$, $A_V = 0.8$

Description: Naming convention: "ins/ta/wat/ins". For example, a storage wall made of 1-cm thick tinted acrylic, a 10-cm thick water wall, and 5-cm thick insulation layers on the inner and outer sides of the storage wall is denoted as ins(5 cm)/ta(1 cm)/wat(10 cm)/ins(5 cm). The thicknesses of the simulated water walls vary from 10 cm to 40 cm with increments of 10 cm. The thickness of the tinted acrylic is 1 cm and the thickness

Table 6. *(Continued)*

of the insulation layers is 5 cm on the inner and outer sides.

Case 7: Water with tinted acrylic at its center

Material: Water and tinted acrylic

Surface properties: Water*: $A_T = 0.9$, $A_S = 0.8$, $A_V = 0.8$

Tinted acrylic: $A_T = 0.9$, $A_S = 0.8$, $A_V = 0.8$

Description: Naming convention: "wat/ta/wat". For example, a storage wall made of 10-cm thick water walls with a 1-cm thick tinted acrylic layer in the middle is denoted as wat(10 cm)/ta(1 cm)/wat(10 cm). Also, "OW" and "IW" are used specify the inner and outer parts of the storage wall. For example, "OW:wat/ta/wat" is used to refer to the thermal energy stored in the outer water wall. The total thickness of the storage wall varies from 11 cm to 41 cm with increments of 10 cm. For all cases considered the thickness of the tinted acrylic is 1 cm and the thicknesses of the outer and inner water walls are equal.

Case 8: Water with tinted acrylic at its center and insulation at its inner side

Material: Water, tinted acrylic and aerogels

Surface properties: Water*: $A_T = 0.9$, $A_S = 0.8$, $A_V = 0.8$

Tinted acrylic: $A_T = 0.9$, $A_S = 0.8$, $A_V = 0.8$

Aerogel: $A_T = 0.9$, $A_S = 0.8$, $A_V = 0.8$

Description: Naming convention:"wat/ta/wat/ins". For example, a storage wall made of 10-cm thick water walls with a 1-cm thick tinted acrylic layer in the middle and a 5-cm thick layer of insulation at its inner side is denoted as wat(10 cm)/ta(1 cm)/wat(10 cm)/ins(5 cm). Also, "OW" and "IW" are used specify the inner and outer parts of the storage wall. For example "OW:wat/ta/wat/ins" is used to refer to the thermal energy stored in the outer water wall. The thicknesses of the simulated storage walls vary from 16 cm to 46 cm with increments of 10 cm. The thicknesses of the tinted acrylic and insulation are always 1 cm and 5 cm, respectively. The thickness of the inner and outer water walls are kept equal.

Table 6. *(Continued)*

Case 9: Water with tinted acrylic at its center and insulation on both sides

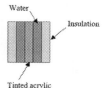

Material: Water, tinted acrylic and aerogels

Surface properties: Water*: $A_T = 0.9$, $A_S = 0.8$, $A_V = 0.8$

Tinted acrylic: $A_T = 0.9$, $A_S = 0.8$, $A_V = 0.8$

Aerogel: $A_T = 0.9$, $A_S = 0.8$, $A_V = 0.8$

Description: Naming convention: "ins/wat/ta/wat/ins". For example, a storage wall made of 10-cm thick water walls with a 1-cm thick tinted acrylic layer in the middle and a 5-cm thick layers of insulation on its inner and outer sides is denoted as ins(5 cm)/wat(10 cm)/ta(1 cm)/wat(10 cm)/ins(5 cm). Also, "OW" and "IW" are used specify the inner and outer parts of the storage wall. For example, "OW:ins/wat/ta/wat/ins" is used to refer to the thermal energy stored in the outer water wall. The thickness of the storage wall varies from 21 cm to 51 cm with increments of 10 cm. For all cases the thickness of the tinted acrylic is 1 cm and the thickness of the insulation on the inner and outer sides of the storage wall is 5 cm. The thickness of the inner and outer water walls varies but is kept equal.

Note: *The properties of the first layer in the partition wall set the absorption for the thermal storage wall in the simulations. When water is the first layer, its absorptive properties are set to that of the absorbing layer within the wall.

can be stored in the water-based Trombe wall and used for building applications as described in Section 4.3 and Section 5, respectively.

4.3 Comparison of Thermal Energy Stored and Water Temperature During Summer for Different Water-Based Trombe Wall Configurations

Fig. 6 shows a comparison of the thermal energy stored in Trombe walls consisting of water storage walls with a total thickness of 20 cm. The simulations used to attain these results were performed using the weather data for July 15, 2020, in Toronto, Ontario. For the bare water storage wall, wat(20 cm), 30.8 MJ of thermal energy is stored. Furthermore, the maximum temperature, T_M, is 29.6°C and the change in water temperature, ΔT, is 3.1°C. For the results shown in this work the maximum temperature occurred at 2 p.m. for all cases (the change in temperature, ΔT, which is provided for all cases shown in Fig. 6, is the increase in temperature between sunrise and 2 p.m.). When tinted acrylic is placed at the

Table 7. Model Options Selected in Design Builder.

Model Options	Configurations Available	Configuration Selected
Data tab		
HVAC	• Simple • Detailed	Simple
Natural ventilation	• Scheduled • Calculated	Calculated
Heating design tab		
Temperature control	• Air temperature • Operative temperature • Radiant temperature	Air temperature
Simulation tab		
Period	• Simulation periods start and end the day	For Summer: July 14 – July 16 For winter: Feb 2 – Feb. 4
Time steps per hour	• 2,4,6,10,14,30,60	4
Temperature control	• Air temperature • Operative temperature	Air temperature
Solar distribution	• Minimal shadowing • Full interior • Full interior and exterior	Full interior and exterior
Solution algorithm	• Conduction transfer function • Finite difference	Conduction transfer function
Finite difference scheme	• Fully implicit first order • Crank–Nicholson second order	Fully implicit first order
Inside convection algorithm	• Adaptive convection • Simple • CIBSE • TARP	TARP
Outside convection algorithm	• Adaptive convection • Simple combined • CIBSE • TARP • DOE-2 • MoWiTT	DOE-2

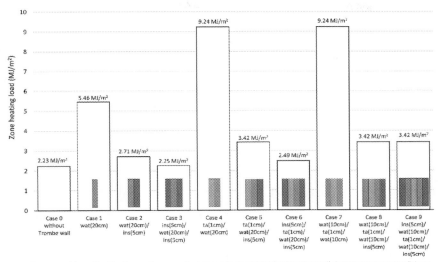

Cases considered for the thermal energy storage medium within the Trombe wall during summer conditions

Fig. 5. Comparison of Total Cooling Load in Different Water Trombe Wall With 20 cm Thickness During Summer on July 14, July 15, and July 16, 2020, in Toronto, Ontario.

outer surface of the water wall (to form ta(1 cm)/wat(20 cm)), the amount of thermal energy stored increases to 111.7 MJ, ΔT is 11.2°C, and T_M is 37.7°C. When the 5-cm thick insulation layer is placed at the inner surface of the water wall, to form wat(20 cm)/ins(5 cm), the amount of thermal energy stored increases to 111.6 MJ, ΔT is 11.2°C, and T_M is 37.7°C. When tinted acrylic is placed at the outer surface of the water wall and insulation is placed at its inner surface, to from ta(1 cm)/wat(20 cm)/ins(5 cm), the amount of thermal energy stored increases to 279.7 MJ, ΔT increases to 28.1°C, and T_M increases to 54.6°C. When insulation is placed on either side of the water wall, to form ins(5 cm)/wat(20 cm)/ins(5 cm), the amount of thermal energy stored is 137.7 MJ, ΔT is 13.8°C, and T_M is 40.3°C.

When the water wall has tinted acrylic at its outer surface and insulation on both surfaces, as in the case of the ins(5 cm)/ta(1 cm)/wat(20 cm)/ins(5 cm) sample, the amount of thermal energy stored increases to 302 MJ, ΔT increases to 30.3°C, and T_M increases to 56.8°C. If the tinted acrylic is placed at the mid-plane of the water storage medium rather than its outer surface, to form ins(5 cm)/wat(10 cm)/ta(1 cm)/wat(10 cm)/ins(5 cm), T_M for the outer portion of the water wall increases slightly to 57.7°C, although T_M for the inner portion of the water wall slightly decreases to 56.1°C. The total thermal energy stored for ins(5 cm)/wat(10 cm)/ta(1 cm)/wat(10 cm)/ins(5 cm) is 302.6 MJ.

It is difficult to compare the results from this chapter with the cooling loads and thermal energy stored in Trombe walls reported in the literature due to differences in

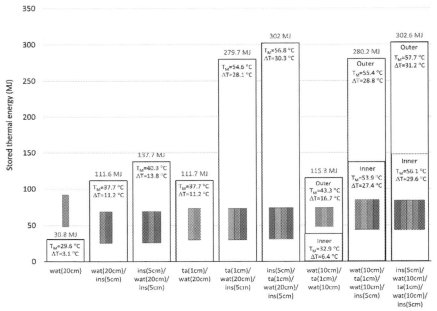

Fig. 6. Comparison of Thermal Energy Stored in Different Trombe Wall Models with 20-cm Wall Thickness During Summer on July 15th, 2020, in Toronto, Ontario ("Outer" and "Inner" Refer to the Outer and Inner Parts of the Thermal Energy Storage Medium, Respectively), ("Wat" = Water, T_M = Maximum Temperature (°C), ΔT = Maximum Increase in the Temperature of the Storage Wall (°C)).

the study parameters including the building structure, geographical location, and weather conditions, although rough comparisons can be made. Mohamad et al. (2019) calculated that water temperatures in excess of 40°C for a Trombe wall with a 0.15-m thick water-storage tank with a black absorber plate on its wall which was subjected to incident sunlight in a Trombe wall configuration. In comparison, as shown in Fig. 6, a maximum temperature of T_M = 37.7°C was achieved when the storage medium of the Trombe wall was a 0.2-m thick water wall with tinted glass at its front surface without insulation (ta(1 cm)/wat(20 cm)) which shows a comparable configuration and results to the work reported by Mohamad et al. Furthermore, Singh and O'Brien (2022) conducted experiments on a small lab-scale water-based Trombe wall prototype to measure the temperature of water within a 0.2-m thick thermal storage medium. When a tinted acrylic sheet was placed in the water storage medium to absorb incident light from a solar simulator, the temperature at the top and bottom of the water storage medium increased to ~55°C and ~34°C, respectively. These results are fairly comparable to those reported in Fig. 6. Future experimental work on full-scale water-based Trombe walls is still required to verify the concept of using water-based Trombe walls for heating water for building applications.

5. Discussion

The heated water in the water-based Trombe walls can be used for applications other than space heating during summer weather conditions. For example, it has been observed that dishwashers operate efficiently at temperatures as low as 38°C, and comfortable temperatures for showering are between 37 and 40°C. Furthermore, the recommended temperature in newer washing machines for high efficiency is about 49°C. The safest recommended temperature for a hot water tank is 49°C, and the inlet water temperature in the water tank during summer is approximately 18°C. Thus, inlet water can be preheated in a water-based Trombe wall to the desired temperature which will help reduce water heating costs and energy consumption (Parachute).

The temperature of the water within the Trombe wall is approximately 57°C for some cases reported in this work. This hot water can be extracted from the top side of the Trombe wall by an outlet valve to use for showering, washing clothes and dishes, and providing preheated water for the hot water tank. As an example, in Canada, on average 75 L/day of hot water are used per person and 255 L/day are used by households. The primary uses of hot water are in showers (~25%), faucets (~34%), baths (~17%), laundry (~15%), and dishwashers (~4%) as shown in Table 8.

Notably, the volume of water in the thermal energy storage medium (for the case of a 20-cm thick water wall) in the single-story building modeled in Design Builder (Fig. 2) is 2.4 m³ (2,400 L). If water was extracted from the top third of the water wall, this would amount to 800 L of heated water on a sunny day that could be used for the domestic purposes described in Table 8. Moreover, there would be excess heated water as 800 L is 3.5 times more than the required amount (255 L) of heated water in an average household in Canada (Natural Resources Canada, 2012; Water Heaters).

Further work is required to build and test full-scale Trombe walls and to experimentally validate their performance including proper water filtration and

Table 8. Main Uses of Hot Water in an Average Household in Canada.

Main Uses of Hot Water	Hot Water Usage (%)	Volume of Hot Water Used (L)
Showers	25	56.25
Faucets	34	76.5
Bath	17	38.25
Laundry	15	33.75
Dishwashers	4	9

quality for targeted applications. The water-based Trombe walls are expected to be the most beneficial in countries that experience hot weather conditions for long periods throughout the year. Further simulations and experimental work needs to be carried out to conduct techno-enviro-economic assessments for water-based Trombe walls in different geographical locations.

6. Conclusions

In this work the potential to use water-based Trombe walls to provide heated water for building applications during the summer months was investigated. Design Builder software was used to model a simple single-story building with a south-facing Trombe wall for winter and summer weather conditions in Toronto, Canada. The effects of using different thermal storage mediums within the Trombe wall on building heating loads during the winter and building cooling loads during the summer were modeled. The amount of thermal energy stored and temperature of water within the thermal storage medium during hot weather conditions were also simulated. The results suggest that water-based Trombe walls have great potential to enhance the flexibility and utility of Trombe walls by providing heated water for building applications during summer months, without compromising performance during winter months. On a sunny day during the summer in Toronto, Canada, the average temperature of the water in a Trombe wall integrated into a single-story building is ~57°C, which is high enough to provide for the main hot water usages in buildings. Furthermore, the amount of water heated is three times greater than that required in an average household in Canada.

References

Abergel, T., & Delmastro, C. (2020). *Is cooling the future of heating?* International Energy Agency (IEA). https://www.iea.org/commentaries/is-cooling-the-future-of-heating

Adams, S., Becker, M., Krauss, D., & Gilman, C. (2010). Not a dry subject: Optimizing water Trombe wall. *academia.edu*. [Online]. https://www.academia.edu/download/52858625/548.pdf. Accessed on October 17, 2022.

Allen, J. G., & Ibrahim, A. M. (2021). Indoor air changes and potential implications for SARS-CoV-2 transmission. *JAMA, 325*(20), 2112–2113. https://doi.org/10.1001/jama.2021.5053

Baetens, R., Jelle, B. P., & Gustavsen, A. (2011, April). Aerogel insulation for building applications: A state-of-the-art review. *Energy and Buildings, 43*(4), 761–769. https://doi.org/10.1016/J.ENBUILD.2010.12.012

Bourdeau, L. (1980). *Study of two passive solar systems containing phase change materials for thermal storage.* Los Alamos National Laboratory.

Chen, W., & Liu, W. (2008, August). Numerical analysis of heat transfer in a passive solar composite wall with porous absorber. *Applied Thermal Engineering, 28*(11–12), 1251–1258. https://doi.org/10.1016/J.APPLTHERMALENG.2007.10.017

Chen, W., & Liu, W. (2004, June). Numerical analysis of heat transfer in a composite wall solar-collector system with a porous absorber. *Applied Energy, 78*(2), 137–149. https://doi.org/10.1016/J.APENERGY.2003.07.003

De Gracia, A., Navarro, L., Castell, A., Ruiz-Pardo, Á., Álvarez, S., & Cabeza, L. F. (2012, January). Solar absorption in a ventilated facade with PCM. Experimental results. *Energy Procedia, 30*, 986–994. https://doi.org/10.1016/J.EGYPRO.2012.11. 111

Energy Plus. https://energyplus.net/. Accessed on November 2, 2022.

Government of Canada. Water temperature and burns/scalds – Canada.ca. https:// www.canada.ca/en/public-health/services/water-temperature-burns-scalds.html. Accessed on October 17, 2022.

Hu, Z., He, W., Ji, J., & Zhang, S. (2017, April). A review on the application of Trombe wall system in buildings. *Renewable and Sustainable Energy Reviews, 70*, 976–987. https://doi.org/10.1016/J.RSER.2016.12.003

IEA. (2023). https://www.iea.org/data-and-statistics/data-tools/energy-efficiency-indicators-data-explorer. Accessed on February 10, 2023.

Kaushik, S. C., & Kaul, S. (1989, January). Thermal comfort in buildings through a mixed water-mass thermal storage wall. *Building and Environment, 24*(3), 199–207. https://doi.org/10.1016/0360-1323(89)90033-4

Leang, E., Tittelein, P., Zalewski, L., & Lassue, S. (2017, September). Numerical study of a composite Trombe solar wall integrating microencapsulated PCM. *Energy Procedia, 122*, 1009–1014. https://doi.org/10.1016/J.EGYPRO.2017.07.467

Long, J., Yongga, A., & Sun, H. (2018, July). Thermal insulation performance of a Trombe wall combined with collector and reflection layer in hot summer and cold winter zone. *Energy and Buildings, 171*, 144–154. https://doi.org/10.1016/ J.ENBUILD.2018.04.035

Mohamad, A., Taler, J., & Ocłoń, P. (2019, January). Trombe wall utilization for cold and hot climate conditions. *Energies, 12*(2), 285. https://doi.org/10.3390/ EN12020285

Natural Resources Canada. (2012). Water heater guide. Google Search. https://www. google.com/search?q=Natural+Resources+Canada.+(2012). +Water+Heater+Guide.&oq=Natural+Resources+Canada.+(2012). +Water+Heater+Guide.&aqs=chrome.69i57j33i16012. 1255j0j15&sourceid=chrome&ie=UTF-8. Accessed on October 17, 2022.

Nayak, J. K. (1987, January). Thermal performance of a water wall. *Building and Environment, 22*(1), 83–90. https://doi.org/10.1016/0360-1323(87)90045-X

Omara, A. A. M., & Abuelnuor, A. A. A. (2020, December). Trombe walls with phase change materials: A review. *Energy Storage, 2*(6), e123. https://doi.org/10.1002/ EST2.123

Ozin, G. A., Kherani, N. P., O'Brien, P. G., Mahtani, P., Chutinan, A., & Leong, K. (2011, August). Selectively transparent and conducting photonic crystal rear-contacts for thin-film silicon-based building integrated photovoltaics. *Optics Express, 19*(18), 17040–17052. https://doi.org/10.1364/OE.19.017040

Parachute. Hot tap water. https://parachute.ca/en/injury-topic/burns-and-scalds/hot-tap-water/. Accessed on October 17, 2022.

Pittaluga, M. (2013, November). The electrochromic wall. *Energy and Buildings, 66*, 49–56. https://doi.org/10.1016/J.ENBUILD.2013.07.028

Rodriguez-Ubinas, E., Montero, C., Porteros, M., Vega, S., Navarro, I., Castillo-Cagigal, M., Matallanad, E., & Gutierrez, A. (2014, November). Passive design strategies and performance of net energy plus houses. *Energy and Buildings, 83*, 10–22. https://doi.org/10.1016/J.ENBUILD.2014.03.074

Saadatian, O., Sopian, K., Lim, C. H., Asim, N., & Sulaiman, M. Y. (2012, October). Trombe walls: A review of opportunities and challenges in research and development. *Renewable and Sustainable Energy Reviews, 16*(8), 6340–6351. https://doi.org/10.1016/J.RSER.2012.06.032

Salari, M., & Javid, R. J. (2016, November). Residential energy demand in the United States: Analysis using static and dynamic approaches. *Energy Policy, 98*, 637–649. https://doi.org/10.1016/j.enpol.2016.09.041

Schultz, J. M., & Jensen, K. I. (2008, March). Evacuated aerogel glazings. *Vacuum, 82*(7), 723–729. https://doi.org/10.1016/J.VACUUM.2007.10.019

Singh, H., & O'Brien, P. G. (2022, October). Semi-transparent water-based Trombe walls for passive air and water heating. *Build, 12*(10), 1632. https://doi.org/10.3390/BUILDINGS12101632

Solarcomponents. (2019). http://www.solar-components.com/TUBES.HTM. Accessed on October 17, 2022.

Team DesignBuilder. (2019). DesignBuilder v6 simulation. Google Scholar. https://scholar.google.com/scholar?hl=en&as_sdt=0%2C5&q=Team+DesignBuilder+%282019%29.+DesignBuilder+v6+Simulation+Documentation.&btnG=. Accessed on October 17, 2022.

Turner, R. H., Liu, G., Cengel, Y. A., & Harris, C. P. (1994, November). Thermal storage in the walls of a solar house. *Journal of Solar Energy Engineering, 116*(4), 183–193. https://doi.org/10.1115/1.2930080

Upadhya, M., Tiwari, G. N., & Rai, S. N. (1991, January). Optimum distribution of water-wall thickness in a transwall. *Energy and Buildings, 17*(2), 97–102. https://doi.org/10.1016/0378-7788(91)90002-K

United Nations Climate Change. *The Paris agreement.* What is the Paris agreement? https://unfccc.int/process-and-meetings/the-paris-agreement/the-paris-agreement. Accessed on November 10, 2022.

United Nations Environment Programme. (2021). *2021 global status report for buildings and construction: Towards a zero-emission.* https://www.unep.org/resources/report/2021-global-status-report-buildings-and-construction

Wang, D., Hu, L., Du, H., Liu, Y., Huang, J., Xu, Y., & Liu, J. (2020, May). Classification, experimental assessment, modeling methods and evaluation metrics of Trombe walls. *Renewable and Sustainable Energy Reviews, 124*, 109772. https://doi.org/10.1016/J.RSER.2020.109772

Wang, W., Tian, Z., & Ding, Y. (2013, September). Investigation on the influencing factors of energy consumption and thermal comfort for a passive solar house with water thermal storage wall. *Energy and Buildings, 64*, 218–223. https://doi.org/10.1016/J.ENBUILD.2013.05.007

Water Heaters. https://www.nrcan.gc.ca/energy-efficiency/products/water-heaters/13735. Accessed on October 17, 2022.

Yang, Y., O'Brien, P. G., Ozin, G. A., & Kherani, N. P. (2013, November). See-through amorphous silicon solar cells with selectively transparent and conducting photonic crystal back reflectors for building integrated photovoltaics. *Applied Physics Letters, 103*(22), 221109. https://doi.org/10.1063/1.4833542

Chapter 5

Investigating Waste Management Efficiencies and Dynamics of the EU Region

Fazıl Gökgöz and Engin Yalçın

Abstract

Waste management is one of the vital objectives for the EU since it has a substantial effect on the environment. European Commission expects annual waste creation on Earth to increase by 70% by 2050. European Commission also estimates that efficient waste management might boost the EU economy's gross domestic product (GDP) by 0.5% by 2030. Hence, it is essential to conduct research including both efficiency and influencing factors analysis for effective waste management. First, we employ both slack-based measure (SBM) and super-SBM data envelopment analysis approaches to investigate the waste management efficiency of the EU region and distinguish between efficient countries. The countries with small areas such as Luxembourg and Ireland have demonstrated super efficiency. Second, we maintain our empirical research with ordinary least square analysis to explore the determinants of waste management. We also conclude that population density, GDP per capita, and tourism rise the amount of waste generated in the EU region.

Keywords: SBM; super-SBM DEA; waste management efficiency; OLS; EU; sustainability

Nomenclature

ADF—Augmented Dickey-Fuller
CE—Circular Economy
DEA—Data Envelopment Analysis
DMUs—Decision-Making Units
DDF—Directional Distance Function

Pragmatic Engineering and Lifestyle, 91–111
Copyright © 2023 Fazıl Gökgöz and Engin Yalçın
Published under exclusive licence by Emerald Publishing Limited
doi:10.1108/978-1-80262-997-220231005

EEA—European Economic Area
EU—European Union
EDA—Explanatory Data Analytics
GDP—Gross Domestic Product
GHG—Greenhouse Gas
GCF—Gross Capital Formation
OLS—Ordinary Least Square
R&D—Research & Development
SBM—Slack-Based Measure
SFA—Stochastic Frontier Analysis
VIF—Variance Inflation Factor
WM—Waste Management

1. Introduction

Today, the world's economic growth and environmental protection have been increasingly polarized; the problem of environmental pollution has become a major impediment to various countries' economic progress. To tackle these problems, countries need to keep in equilibrium economic growth and environmental management (Baloch et al., 2019). The production, disposal, and usage of goods have caused severe problems both for the climate and human health (Shen et al., 2020). In this regard, waste management (WM) is an important contributor to providing this balance as poor WM causes a lack of environmental sustainability and brings about human health problems (Serge Kubanza & Simatele, 2020).

WM is a topic that needs to enhance the economic and environmental aspects of efficiency (Sarra et al., 2020). This issue entails proper disposal, reuse, and recycling of goods (Campitelli & Schebek, 2020). Waste efficiency measurement could enable decision-makers to observe the current status of countries. Thereby, countries can utilize recycled materials and decrease waste amounts as much as possible. WM is also of the most prominent priorities in the European Union (EU).

The environmental policy of the EU stresses WM that is eco-friendly and utilizes secondary resources. The Waste Framework Directive determines the EU's framework for waste treatment and management. It introduces the "waste hierarchy," an order for WM. The EU waste strategy depends on protecting the climate and human health simultaneously, also aiding in achieving its circular economy (CE) goals. It lays out objectives for better WM, recycling, and landfill reduction (European Commission, 2022a, 2022b). In 2018, the EU-27 generated an average of 5.2 tons of waste from all sources per person (European Environment Agency, 2022b). This number points out the necessity for efficient management of waste in EU countries. Hence, controlling and recycling waste is at the forefront of the EU to reach long-term targets.

The data envelopment analysis (DEA) is an approach focusing on determining efficient options and the most suitable allocation of resources (Gökgöz, 2010). So far, many versions of the DEA method have been used in environmental literature. The traditional DEA technique has drawbacks since it ignores the slack variables of input and output, leading to substantial variances in efficiency (Keskin, 2021). To this end, the slack-based measure (SBM) approach, introduced by Li and Shi (2014), has addressed this issue, which can help distinguishing each efficiency value, resulting in more precise and efficient allocation results. Instead of using traditional radial DEA models, the SBM and super-SBM techniques can be regarded as more advantageous (Chen et al., 2021). The Super-SBM model differentiates between several decision-making units (DMUs) and includes slacks in the objective function. It is also a model capable of adding undesirable outputs into the process (Yan et al., 2019). As WM is an issue involving undesirable outputs, the super-SBM model can be regarded as an appropriate technique for efficiency measurement. First, we intend to investigate the WM efficiency of the EU via the SBM and super-SBM approaches. Second, we aim to determine the factors that affect waste generation via the ordinary least square (OLS) technique, which is widely employed to explore the relationships between dependent and independent variables. Appropriate measurement of influencing factors in waste generation is a prerequisite to presenting a complete and comprehensive insight into countries. Thereby, the influencing factors of WM can be observed via the OLS technique, which will further enlighten the EU waste structure. The SBM approach has been employed to evaluate energy efficiency on a country basis due to the availability of undesirable output namely, CO_2 emissions. Likewise, WM management is a field that contains undesirable output and is suitable for the SBM approach. Besides, the SBM approach cannot discern efficient units. To overcome this situation, the super-SBM technique was introduced. The original contribution of this study is to analyze WM efficiency from a different perspective for the EU countries. The available studies generally focus on traditional DEA techniques, whereas our study employs a novel technique for WM efficiency namely, SBM and super-SBM DEA techniques, which hold more discriminating power in comparison to the traditional DEA method and treat undesirable outputs and slacks. We also conduct an OLS analysis, which is widely used to determine relationships among variables, to distinguish our study from current studies in WM dynamics for the EU.

Our study proceeds as follows. We present a literature review of WM studies. Then, we introduce and employ efficiency and OLS analysis techniques used in our empirical analysis. The last part of our study elaborates on the conclusion and policy implications.

2. Literature Review

So far, many papers have investigated the WM efficiency and dynamics of countries by various decision-making techniques. We briefly include these studies as follows:

Halkos and Petrou (2019) investigated the WM in the European region via the DEA and directional distance function. They include novel techniques and parameters for WM efficiency. Their indicators include waste generated by households, gross domestic product (GDP), number of employees, gross capital formation (GCF), population density, and emissions from the waste sector. They conclude that the top efficiency ratings belong to countries, which possess the highest recycling rates, such as Germany, Ireland, and the United Kingdom. Besides, countries that use treatment methods with high use of more sustainable ones have also proven to be efficient.

Taboada et al. (2020) investigated the WM in the European Economic Area (EEA) by the DEA and explanatory data analytics (EDA). Their efficiency indicators comprise 15 criteria, which are the components of construction and demolition in WM. The initial part comprises an EDA analysis of sustainability in the EEA. Then, they proceed with the DEA to evaluate waste efficiency with an enhanced and novel measurement model, which aims to overcome the deficiencies of the current criteria. They infer that the suggested DEA criteria have proven to be a useful option to differentiate efficiency between countries.

Yang et al. (2018) evaluated the WM in the Chinese provinces by the DEA and stochastic frontier analysis (SFA). The fuzzy c-means algorithm is employed for cluster research of the Chinese provinces, as per DEA results. They select equipment and fixed asset investment as inputs, while collected solid waste and solid waste treatment rates are used as outputs. They conclude that WM efficiency is low in general and is projected to improve. They infer that WM efficiency has been demonstrated to vary by region.

Giannakitsidou et al. (2020) investigated environmental efficiency in the European countries depending on waste generation by the DEA. They aim to comprise a holistic evaluation framework based on waste per capita. They form two models to evaluate environmental and CE performance. Their environmental model includes generated solid waste, human needs, well-being foundation, and opportunity as inputs; the recycling rate of solid waste is used as output, while the inputs of the CE performance model are the same as the previous model. Circular material use rate is included as an output in addition to the recycling rate. They conclude that significant performance disparities exist among the investigated countries. They discover that Spain and France performed remarkably poorly in comparison to the latest member countries.

Agovino et al. (2018) analyzed the WM in the Italian provinces by the DEA. They employ urban and unsorted waste as outputs, while the institutional quality index, value-added per capita, and some socioeconomic factors such as population density, unemployment rate, and consumption per capita are used as inputs. Their analysis results present that northern and southern cities performed well while the performance of central ones remained quite low. The efficiency scores are then found to have a positive geographical connection. They infer that when citizens and local governments act responsibly, the WM process improves.

Díaz-Villavicencio et al. (2017) evaluated the WM dynamics of some Spanish municipalities via directional distance function (DDF) and Tobit regression. They include population, glass, paper, cardboard, light packaging, and other waste as technical criteria. Then, a relational analysis is conducted with socioeconomic criteria. Tourism and education were significant indicators as per Tobit approach. The analysis results reveal that a government does not require to possess an excessive economic development or be densely populated to achieve optimal WM efficiency.

Bartolacci et al. (2019) assessed the effect of waste on scale economies in Italian WM companies. They include the number of employees, waste collected quantity, total costs, waste collection rate, and population density as explanatory variables. They estimate this relationship by employing the OLS method. As per the analysis results, an increase in solid waste quantity produces an increase in total costs, and the scale economy is affected by population density as well.

Gardiner and Hajek (2020) investigated the causality between waste production and economic growth in the EU region via the Granger causality model. They utilize the panel vector error correction model to reflect this causality. Their indicators include GDP per capita, waste production, GCF, employment rate, and heating days from 2000 to 2018. The analysis results confirm short- and long-term causality between these indicators. They ascertain that a bidirectional causality exists between energy, research & development (R&D) intensity, and waste production.

lo Storto (2021) investigated both the efficiency and dynamics of WM in Italian municipalities by the DEA and Tobit regression. They determine the solid waste cost as input. They select plastic, metal, paper, glass, and other material waste collection as desirable output, while unsorted waste is used as undesirable output to measure efficiency. Then, lo Storto (2021) proceeds with the Tobit analysis to analyze determinants of WM. To use the Tobit regression analysis, sorted waste collection, population, length of urban roads, area of the municipal territory, and average generated waste are used to explore the determinants of WM in Italy. lo Storto (2021) concludes that most Italian municipalities are not close to reaching the efficiency frontier. lo Storto (2021) also infers that increasing the efficiency of WM requires lowering the average amount of waste generated.

3. Research Methodology

3.1 SBM and Super-SBM Models

The DEA is a technique that depends on determining the efficiency of entities (Gökgöz & Erkul, 2019). Its principal feature is to evaluate the efficiency of a unit in comparison to the best DMU based on its deviation from the production frontier. One of the most prominent features of the DEA is that it does not need any functional relationships or a priori weights, thereby enabling objective estimation. The DEA focuses on obtaining more output while employing fewer inputs. During this process, undesirable outputs may exist. So far, traditional DEA models have been implemented in various fields as a successful efficiency technique (Gökgöz & Yalçın, 2022). However, these models are radial and angular, which presume that changes in inputs or outputs are proportionate, and they ignore the availability of slacks (Jiang et al., 2021). In a traditional DEA model, which adopts the feature of being radial, input and output are balanced to the efficiency frontier by the same proportion. Hence, this structure may produce misleading efficiency results (Lee et al., 2020). By contrast with the traditional DEA technique, the SBM approach considers inefficiency from both input and output viewpoints. In comparison to the traditional DEA model, the SBM approach possesses a smoother and more precise frontier (Zhong et al., 2021). Due to its nature of considering slacks, the SBM methodology is regarded to have a more discriminating power in measuring efficiency (Li & Lin, 2017). Tone (2001) introduced the SBM model by adding slacks into the objective function. Then, Tone (2002) enhanced this model with the super-SBM version, which is a nonradial form and combines the SBM and super-efficiency DEA (Zhong et al., 2021). Afterward, Tone (2004) presented the enhanced version of SBM as follows detaching outputs as desirable and undesirable ones (Yan et al., 2019).

$$\min p = \frac{1 - \frac{1}{m}\sum\limits_{i=1}^{m} S_i^- / x_{io}}{1 + \frac{1}{s_1 + s_2}\left(\sum\limits_{r=1}^{s_1} \frac{S_r^g}{y_{ro}^g} + \sum\limits_{r=1}^{s_2} \frac{S_r^b}{y_{ro}^b}\right)} \tag{1}$$

s.t.

$$x_0 = X\lambda + s^-; y_0^g = Y^g\lambda - s^g; y_0^b = Y^b\lambda + s^b \tag{2}$$

$$\lambda \geq 0,\ s^- \geq 0, s^g \geq 0, s^b \geq 0 \tag{3}$$

where n denotes options, all with m inputs and s_1 desirable and s_2 undesirable outputs, λ indicates a positive vector, the vectors s^- and s^b reflect excesses in inputs and undesirable outputs, whereas s^g stands for a lack of desirable output. If the p-value becomes one and the slacks get the value of zero, an alternative is said to be efficient.

The SBM approach can cope with input surpluses and output shortages. This feature renders the SBM technique advantageous over other efficiency techniques. As the slackness is taken into consideration to compose the efficient frontier, the efficiency rating of alternatives becomes more accurate, thanks to the inclusive shape of the frontier directed by slacks (Zhong et al., 2021). The super-SBM model can simultaneously calculate DMU input and, undesirable output redundancy rates, as well as measure DMU efficiency. The formulation of the super-SBM approach is shown below (Long et al., 2020).

$$\min p^* = \frac{1 + \frac{1}{m}\sum\limits_{i=1}^{m} S_i^- / x_{ik}}{\frac{1}{q_1 + q_2}\left(\sum\limits_{r=1}^{q_1} \frac{s_r^{g+}}{y_{rk}} + \sum\limits_{t=1}^{q_2} \frac{s_t^{b-}}{y_{tk}}\right)} \tag{4}$$

s.t.

$$s_i^- \geq \sum\limits_{j=1, j\neq k}^{n} x_{ij}\lambda_j,\ i = 1, \ldots, m; \tag{5}$$

$$s_r^{g+} \leq \sum\limits_{j=1, j\neq k}^{n} y_{rj}^g\lambda_j, r = 1, \ldots, m; \tag{6}$$

$$s_r^{g+} \geq \sum_{j=1j\neq k}^{n} y_{tj}^b \lambda_j, \quad t = 1,\ldots,m; \tag{7}$$

$$\lambda_j \geq 0, j = 1,\ldots,n \quad j\neq0; \tag{8}$$

$$s_i^- \geq x_{ik}, \quad i = 1,\ldots,m; \quad s_r^{g+} \leq y_{rk}, r = 1,\ldots,q_1; \quad s_t^{b-} \geq y_{tk}, \quad t = 1,\ldots,q_2 \tag{9}$$

where p^* denotes efficiency value, s^-, s_r^{g+}, s_t^{b-} signify the input, desirable and undesirable output slacks, respectively while λ_j denotes constraint. If the super-SBM model's objective function value is one, all input and output slacks are zero, and the entity under assessment is found as efficient (Gerami et al., 2021).

3.2 The OLS Analysis

The OLS analysis tries to model the relationship between variables by fitting a linear equation to the data. This statistical analysis method conducts mathematical calculations in order to relate the dependent variable with independent variables (Amid et al., 2017). Two main types of OLS techniques exist based on complexity among variables, namely, simple and multiple OLS approaches, which include one and more than one predictor, respectively (Fumo & Biswas, 2015). The past is assumed to be a surrogate in the future for OLS analysis. As a result, this technique assumes that future predicted values could follow the structure of previous data. The formulation of multiple variable OLS analysis is specified as follows (Hosseini et al., 2019).

$$Y = \beta_0 + \beta_1 X_1 + \beta_2 X_2 + \ldots + \beta_p X_p + \epsilon \tag{10}$$

where Y denotes a dependent variable, signify the independent variables, $\beta_0, \beta_1, \ldots, \beta_p$ OLS coefficients; ϵ denotes the error term that accounts for the difference between estimated and observed data. The single-index OLS approach is used to forecast the variables. By minimizing the sum of the squared differences between the observed dependent variables in data, OLS determines the parameters of a linear function of a collection of independent variables (Barhmi et al., 2020).

4. WM Overview of the EU Countries

The EU's environmental policy emphasizes WM that is environmentally friendly and utilizes the secondary resources it contains. The goal of the EU waste policy is to enhance CE by taking out as many sources as possible from waste and protecting the environment and human health (European Commission, 2022a). The EU determines various objectives to improve WM performance. In this section, we aim to present a brief overview of to what extent the EU countries enhance WM performance.

One of the significant sources of greenhouse gases (GHG) emissions is waste, which jeopardizes the environment and human health. The amount of waste-related emissions is determined by how the waste is treated. When waste is landfilled, for instance, the organic stuff decomposes and creates gas. Less waste has to be landfilled or incinerated because of increased recycling, which helps protect the environment (Eurostat, 2022c).

Fig. 1 illustrates average GHG emissions from the waste in the EU over the investigated period.

As Fig. 1 indicates, average GHG emissions from waste decreased by 9% over the studied period. This figure is a positive sign regarding environmental protection in the EU region.

As mentioned earlier, the increase in recycling rates indicates an enhancement by employing waste as a resource. Thus, it is significant to improve recycled waste amounts (European Environment Agency, 2022a). Fig. 2 presents the average waste recovery amount over the 2012–2018 period.

As can be observed from Fig. 2, waste recovery has an increasing trend. However, only 38% of waste in the EU is recovered, which indicates an improvement probability regarding waste recovery in the EU (Eurostat, 2022b).

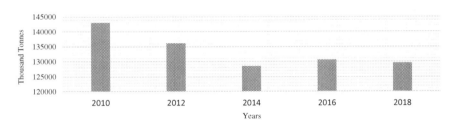

Fig. 1.   Average GHG Emissions From Waste in the EU Countries.

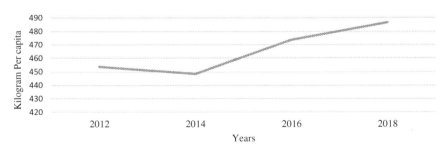

Fig. 2.   The Average Waste Recovery Amount in the EU Countries.

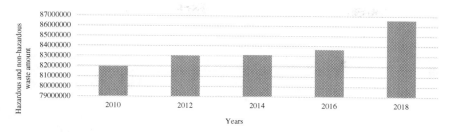

Fig. 3.   The Average Waste Amount in the EU Countries.

The existing production and consumption model named the linear economy generates a great deal of waste that is rarely reused or recycled. The waste produced has a significant influence on the environment and living beings. Fig. 3 illustrates the average waste generation in the EU countries over the studied period.

As can be observed from Fig. 3, the average waste has an increasing trend over the investigated period. In this regard, recycling and reuse of waste has an immense effect both on the environment and the economy. The CE aims to create value by reintroducing consumed materials into the economy. The most efficient way of dealing with these problems is to focus on a CE instead of a linear economy, which intensifies its efforts solely on the production and consumption of goods.

5. Empirical Analysis

5.1 Data and Variables

Our empirical analysis comprises two parts, which aim to present a holistic perspective on the WM efficiency of EU members. First, we intend to clarify WM efficiency in the EU countries by the SBM and super-SBM models. As this technique can include undesirable output, namely, GHG emissions from waste in our analysis, it can be regarded as appropriate to employ this technique. Besides, it is advantageous as efficient countries can be distinguished in this approach. After elaborating on efficiency measurement, we proceed with an econometric investigation, which will further enlighten the WM structure of the EU members. The data used in our research analysis are retrieved from the Eurostat (2022a) databank. Our technical criteria are selected based on available papers on WM efficiency (Callao et al., 2019; Halkos & Papageorgiou, 2016; Halkos & Petrou, 2019) and data availability. The developed countries generate a high amount of waste due to their industry-intensive operations. During these operations, waste generation becomes inevitable. Our empirical model aims to evaluate WM efficiency, namely, obtaining as high as possible economic output, while simultaneously reducing GHG emissions from waste and increasing recovered waste. The slack-based methodology is an appropriate technique to provide the economic output and waste generation balance. The conventional DEA techniques are based on radial assumption. The radial version presumes that both input and output changes realize proportionally and in the same direction. Further, the conventional model cannot include undesirable output in the objective function. The real production processes generally produce undesirable outputs. Calculation of efficiency by the conventional model may result in a biased outcome (Jiang, 2021). Hence, we employ slack-based DEA, which can deal with the input–output slack problems and the effect of oriented models (Xiao et al., 2018). We select GCF, labor productivity, population density, and waste generation as our inputs whereas GDP per capita, and recovered waste are selected as desirable outputs, and GHG emissions from waste are determined as undesirable output for the super-SBM model analysis. These variables are selected as they are closely concerned with waste generation and its efficiency measurement. The description of our criteria is presented in Table 1.

Table 2 summarizes the descriptive statistics of the technical criteria used in this study.

Table 1. The Description of Criteria Used in the Efficiency Measurement.

Indicator Type	Technical Criteria	Description
Inputs	Population density	The number of people living in each unit of area.
	Labor productivity	Real economic output per labor hour.
	GCF	The total value of the gross fixed capital formation, changes in inventories, and acquisitions fewer disposals of valuables for a unit.
	Waste generation	All materials discarded, whether or not they are later recycled or disposed in a landfill.
Desirable outputs	GDP per capita	The monetary value of final goods and services produced in a country in a given period.
	Recovered waste	Any WM operation that diverts a waste material from the waste stream and which results in a certain product with a potential economic or ecological benefit.
Undesirable output	GHG emissions from waste	The total amount of solid waste contributing directly to GHG emissions.

Table 2. The Descriptive Statistics of Efficiency Criteria Over the 2012–2018 Period.

Indicator Type	Technical Criteria	Minimum	Average	Maximum	Standard Deviation
Inputs	Population density	18	174	1,426	226
	Labor productivity	9.2	19.1	35.3	7.8
	GCF	11	20.6	29.6	3.7
	Waste generation	1,703,095	84,098,784	390,280,427	101,266,699
Desirable outputs	GDP	5,978	25,591	81,075	16,429
	Recovered waste	222	811	2,817	578
Undesirable outputs	GHG emissions from waste	2,509	131,223	877,824	191,327

Source: Authors' Calculations.

5.2 Analysis Results

We proceed with empirical analysis results to present a detailed insight into the WM efficiency of the EU countries. Table 3 shows the analysis results of the SBM and super-SBM models for the EU members over the 2012–2018 period.

The efficiency scores of the SBM model results vary from 0 to 1. When there are more than one efficient DMUs, the SBM model remains inadequate. Hence, the super-SBM model is used to obtain an efficiency score higher than one for efficient countries, thereby discriminating among efficient units. Therefore, the efficiency scores of inefficient countries remain the same both for SBM and super-SBM models. The super-SBM technique is a synthesis of super efficiency and SBM techniques. As per the empirical results of the SBM model. Belgium, Ireland, Luxembourg, Slovenia, Finland, and Sweden are found efficient as per the SBM result. The SBM method does not reflect which efficient countries perform relatively better. Hence, it is essential to conduct a super-SBM analysis to distinguish between these efficient countries. As per the super-SBM model, Ireland and Luxembourg have attained a better efficiency level in comparison to the remaining EU countries. Bulgaria and Romania are the countries with the lowest efficiency levels regarding waste efficiency. Likewise, developed members of the EU such as Germany, France, and Italy performed far from being effective due to the high amount of waste. Besides, nearly all EU members displayed stable performance over the observed period except Malta, which recorded a critical

Table 3. The SBM and Super-SBM Analysis Results of the EU Countries.

Countries	2012 (SBM)	2012 (Super-SBM)	Rank	2014 (SBM)	2014 (Super-SBM)	Rank	(2016 SBM)	2016 (Super-SBM)	Rank	2018 (SBM)	2018 (Super-SBM)	Rank
Belgium	1.000	1.137	5	1.000	1.250	4	1.000	1.161	3	1.000	1.238	3
Bulgaria	0.110	0.110	23	0.146	0.146	23	0.199	0.199	21	0.199	0.199	22
Czechia	0.231	0.231	20	0.294	0.294	21	0.265	0.265	18	0.268	0.268	20
Denmark	0.599	0.599	9	0.742	0.742	10	0.627	0.627	10	0.644	0.644	9
Germany	0.312	0.312	13	0.416	0.416	13	N/A	N/A	N/A	0.321	0.321	15
Estonia	1.000	1.508	2	1.000	1.610	1	N/A	N/A	N/A	N/A	N/A	N/A
Ireland	1.000	1.113	6	1.000	1.151	6	1.000	1.273	2	1.000	1.326	2
Spain	0.301	0.301	14	0.389	0.389	17	0.361	0.361	13	0.390	0.390	12
France	0.267	0.267	17	0.407	0.407	14	0.433	0.433	12	0.443	0.443	10
Croatia	0.258	0.258	19	0.382	0.382	18	0.349	0.349	14	0.401	0.401	11
Italy	0.297	0.297	15	0.395	0.395	15	0.312	0.312	16	0.317	0.317	16
Lithuania	0.387	0.387	12	0.576	0.576	12	1.000	1.107	6	1.038	1.038	7
Luxembourg	1.000	1.637	1	1.000	1.547	2	1.000	1.697	1	1.000	1.771	1
Hungary	0.194	0.194	21	0.238	0.238	22	0.230	0.230	20	0.242	0.242	21
Malta	1.000	1.088	7	1.000	1.003	9	1.000	1.132	4	0.313	0.313	17
Netherlands	0.446	0.446	11	1.000	1.020	8	0.471	0.471	11	0.295	0.295	18
Austria	0.523	0.523	10	0.603	0.603	11	0.916	0.916	9	0.689	0.689	8
Poland	0.192	0.192	22	0.307	0.307	19	0.246	0.246	19	0.285	0.285	19

Portugal	0.289	0.289	16	0.393	0.393	16	0.315	0.315	15	0.338	0.338	13
Romania	0.089	0.089	24	0.126	0.126	24	0.148	0.148	22	0.172	0.172	23
Slovenia	1.000	1.141	4	1.000	1.307	3	1.000	1.132	5	1.000	1.168	4
Slovakia	0.260	0.260	18	0.307	0.307	20	0.303	0.303	17	0.324	0.324	14
Finland	1.000	1.458	3	1.000	1.185	5	1.000	1.036	8	1.000	1.063	5
Sweden	1.000	1.048	8	1.000	1.031	7	1.000	1.085	7	1.000	1.051	6

Source: Authors' Calculations.

decline in efficiency score in 2018, mostly due to increasing population density and waste generation amount.

In the second stage of our analysis, we proceed with analyzing the WM dynamics by an OLS model in order to present a detailed insight into the EU countries. Before implementing the OLS technique, we need to implement stationary and normality tests to provide the suitability of variables. The normality evaluation of variables is a precondition for the OLS method. Further, the variables used need to be stationary to prevent the problem of pseudorelationship (Yan et al., 2020). Hence, we check the stationary and normality of our variables. We present the results of the Augmented Dickey-Fuller (ADF) and Jarque–Bera normality tests in Table 4.

The results of the ADF test indicate that variables are stationary at 1% significance level. Further, our variables are normally distributed in log form as *p*-values of the Jarque–Bera test are greater than 0.05. After confirming the normality and stationary of our variables, we establish a multiple OLS model as specified below based on studies investigating the waste generation determinants (Cheng et al., 2020; Estay-Ossandon et al., 2018; Zambrano-Monserrate et al., 2021).

$$MSW = \beta_0 + \beta_1 \log(\text{population}) + \beta_2 \log(\text{GDP}) + \beta_3 \log(\text{tourism}) + \mu_{ct} \quad (11)$$

where MSW denotes municipal solid waste generation amount, GDP is gross domestic product per capita, tourism signifies the total number of arrivals at tourist accommodation establishments. μ is the error term. All parameters are defined in log form to reduce the variance of the model error. Table 5 presents the OLS analysis results formed to explore waste determinants in the EU region.

As can be observed from Table 3, our empirical model is found statistically significant based on the *p*-value of F Statistics, which indicates a linear relationship between variables. The R^2 value of our model indicates that about 88% of waste generation variation is explained by variations in our independent variables. Besides, the VIF value is less than 10; we may conclude that there exists no multicollinearity problem among independent variables. As per OLS analysis results, a 1% increase in population density causes a 0.63% increase in generated waste amount. A 1% increase in tourism operations and GDP per capita are also found to increase waste generation in the EU region by 0.45% and 1.23% respectively, as well. Our OLS analysis results are in line with previous studies in this field, which explore waste generation determinants (Arbulú et al., 2015; Namlis & Komilis, 2019; Zambrano-Monserrate et al., 2021). An accurate analysis of waste determinants is a significant part of WM. The positive relationship between GDP and waste production indicates that economic development is likely to increase waste amounts in the EU countries. As per analysis results, GDP per capita is found to be the most influential factor among investigated variables. Hence, EU countries should focus on clean technologies to perform sustainable economic development. Likewise, waste generation increases in highly populated and touristy areas. This result calls for particular policies and codes such as waste reduction and technological improvement systems in these regions.

Table 4. The Results of the ADF and Jarque-Bera Normality Test.

Normality and Stationarity Test Statistics	Log(GDP)	Log(Population)	Log(Tourism)
ADF test value	−3.43	−4.04	−6.00
ADF *p* value	0.00	0.00	0.00
Jarque-Bera statistics	0.43	3.34	0.31
Jarque-Bera *p* value	0.80	0.18	0.85

Source: Authors' Calculations.

6. Discussion of Results

In this chapter, we initially investigate the WM in the EU countries by SBM and super-SBM techniques that deal with the handicap of the traditional DEA model stemming from its radial form. Second, we aim to determine the factors that affect waste generation via the OLS technique to further enlighten the waste structure of the EU.

As for the SBM analysis results, Belgium, Ireland, Luxembourg, Slovenia, Finland, and Sweden are efficient over the analysis period. Luxembourg, Ireland, and Estonia are the countries with the highest level of efficiency regarding the super-SBM model. Furthermore, Eastern members of the EU, such as Bulgaria and Romania remained inefficient compared with other EU members. Among the EU members, Ireland is the country with the maximum recycling rate target. The Irish government obliged the country to obtain a high-level recycling rate within 15 years. Laws and regulations are of fundamental importance to achieving a high level of recycling rates (European Commission, 2015; Li et al., 2020). Luxembourg and Ireland have recycling rates of over 90%, which significantly enhance their WM efficiency (Zhang et al., 2022). Likewise, Scandinavian

Table 5. The OLS Analysis Results of the Waste Determinants in the EU Region.

Parameters	Coefficients	*p* Values
Constant	0.057	0.888
Log(GDP)	1.235	0.040**
Log(Population)	0.639	0.008***
Log(Tourism)	0.450	0.074***
F statistics	18.30	0.000*
Variance inflation factor (VIF)	5.555	
R^2	0.884	

Source: Authors' calculations.

Note: *, **, *** denotes 1%, 5%, and 10% significance level, respectively.

countries, such as Finland and Sweden, are pioneer countries with their waste efficiency due to their steadily increasing recycling rates. Even before the waste legislation, these countries had a tradition of recycling. They have also implemented landfill bans and taxes since the 2000s, which makes them more responsible for WM policies (Salmenperä, 2021). Further, Belgium is a leading country in terms of waste efficiency. The recycling rate remains above the EU average and waste generation amounts below the EU average in Belgium (Joseph et al., 2018).

Landfilling fees and waste taxes implemented by governments can be regarded as suitable options to reuse waste in the economy. Estonia is of the most successful countries in preventing its waste from landfilling, which is mainly due to energy recovery (Malinauskaite et al., 2017). We also may conclude that the smallest size EU members of the EU such as Luxembourg and Estonia perform better due to their area and population advantage as well. A low population causes less waste generation and renders WM more controllable (Pais-Magalhães et al., 2021). In addition, northern countries such as Finland and Sweden have been efficient. Strict national and municipal WM plans, landfill bans, and taxes occupy a key part of their WM efficiency (Salmenperä, 2021).

Then, we conduct an OLS analysis to explore the determinants of the EU. We form an econometric analysis and evaluate the factors that have affected the waste amount via the OLS technique. As per OLS results, we observe that increasing population, economic operations, and tourism are the factors that generate the increase in waste generation. In particular, the operations that cause population increase and economic growth also increase the amount of waste, mostly due to tourism, urbanization, and increasing energy consumption. Recycling waste not only decreases waste generation but also GHG emissions (Ramachandra et al., 2018). These analyses results point out the role of WM in combatting climate change. Efficient WM is of utmost importance specifically for developed countries as they generate a high amount of waste because of their economic operations and relatively high population in comparison to the remaining EU members. Recycling and the balance between economic growth and environmental conservation strategies should be at the forefront of developed EU countries. Further, tourism is a factor that puts pressure on waste generation. Tourism is a significant sector that contributes to economic growth. However, waste production is at very high levels in areas with heavy tourist arrivals. Municipalities have a prominent place in these areas in preventing the increase in waste generation. Food waste decrease, reduction of single-use plastics, and an increase in separate collection and recycling are important precautions in touristy regions (Obersteiner et al., 2021).

7. Conclusion

Today, WM is one of the most significant aims of the EU because of its adverse impacts on the environment and human health. In this regard, efficient use of materials contributes significantly to the endeavors of WM, which will in the end enhance the sustainability of countries. Increasing population and industrialization contain the problem of waste, which is one of the critical threats to both the environment and human health. Hence, developed countries such as EU members attach great importance to WM as efficient management of waste also contributes to the economy of countries. Efficient WM heavily depends on the reuse and recycling of resources, which are the principal components of a CE. Thus, it is highly significant to determine and evaluate to what extent countries get closer to predetermined aims. In this regard, the super-SBM model has a crucial role in observing efficiency levels. Belgium, Ireland, Luxembourg, Slovenia, Finland, and Sweden are found to be efficient throughout the analysis period. Significant alternatives exist for WM efficiency in countries. Zero-waste-oriented policies, namely recycling and reuse of waste, are effective tools to enhance WM efficiency (Palafox-Alcantar et al., 2020).

Then, we conduct an OLS analysis to explore the determinants of the EU. We form an econometric analysis and evaluate the factors that have affected the waste amount via the OLS technique. As per OLS results, we observe that increasing population, economic operations, and tourism are the factors that generate the increase in waste generation.

Consequently, we may infer that decision techniques, namely the SBM, super-SBM, and OLS techniques, are viable options to evaluate the efficiency and dynamics of countries. This study may provide remarkable insight into the WM of the EU.

References

Agovino, M., D'Uva, M., Garofalo, A., & Marchesano, K. (2018). Waste management performance in Italian provinces: Efficiency and spatial effects of local governments and citizen action. *Ecological Indicators, 89*, 680–695.

Amid, S., & Mesri Gundoshmian, T. (2017). Prediction of output energies for broiler production using linear regression, ANN (MLP, RBF), and ANFIS models. *Environmental Progress & Sustainable Energy, 36*(2), 577–585.

Arbulú, I., Lozano, J., & Rey-Maquieira, J. (2015). Tourism and solid waste generation in Europe: A panel data assessment of the environmental Kuznets Curve. *Waste Management, 46*, 628–636.

Baloch, M. A., Mahmood, N., & Zhang, J. W. (2019). Effect of natural resources, renewable energy and economic development on CO_2 emissions in BRICS countries. *Science of the Total Environment, 678*, 632–638.

Barhmi, S., Elfatni, O., & Belhaj, I. (2020). Forecasting of wind speed using multiple linear regression and artificial neural networks. *Energy Systems, 11*(4), 935–946.

Bartolacci, F., Del Gobbo, R., Paolini, A., & Soverchia, M. (2019). Efficiency in waste management companies: A proposal to assess scale economies. *Resources, Conservation and Recycling, 148*, 124–131.

Callao, C., Martinez-Nuñez, M., & Latorre, M. P. (2019). European Countries: Does common legislation guarantee better hazardous waste performance for European Union member states? *Waste Management, 84*, 147–157.

Campitelli, A., & Schebek, L. (2020). How is the performance of waste management systems assessed globally? A systematic review. *Journal of Cleaner Production, 272*, 122986.

Cheng, J., Shi, F., Yi, J., & Fu, H. (2020). Analysis of the factors that affect the production of municipal solid waste in China. *Journal of Cleaner Production, 259*, 120808.

Chen, F., Zhao, T., Xia, H., Cui, X., & Li, Z. (2021). Allocation of carbon emission quotas in Chinese provinces based on Super-SBM model and ZSG-DEA model. *Clean Technologies and Environmental Policy, 23*(8), 2285–2301.

Díaz-Villavicencio, G., Didonet, S. R., & Dodd, A. (2017). Influencing factors of eco-efficient urban waste management: Evidence from Spanish municipalities. *Journal of Cleaner Production, 164*, 1486–1496.

Estay-Ossandon, C., & Mena-Nieto, A. (2018). Modelling the driving forces of the municipal solid waste generation in touristic islands. A case study of the Balearic Islands (2000–2030). *Waste Management, 75*, 70–81.

European Commission. (2015). *Construction and demolition waste management in Ireland*. https://ec.europa.eu/environment/topics/waste-and-recycling/construction-and-demolition-waste_en. Accessed on March 5, 2022.

European Commission. (2022a). https://ec.europa.eu/environment/topics/waste-and-recycling_en. Accessed on March 5, 2022.

European Commission. (2022b). https://ec.europa.eu/environment/topics/waste-and-recycling/waste-framework-directive_en. Accessed on March 5, 2022.

European Environment Agency. (2022a). https://www.eea.europa.eu/data-and-maps/indicators/waste-recycling-1/assessment-1. Accessed on March 5, 2022.

European Environment Agency. (2022b). https://www.eea.europa.eu/themes/waste. Accessed on March 5, 2022.

Eurostat. (2022a). https://ec.europa.eu/eurostat/data/database. Accessed on March 5, 2022.

Eurostat. (2022b). https://ec.europa.eu/eurostat/web/products-eurostat-news/-/ddn-20220913-1. Accessed on March 5, 2022.

Eurostat. (2022c). https://ec.europa.eu/eurostat/web/products-eurostat-news/-/DDN-20200123-1. Accessed on March 5, 2022.

Fumo, N., & Biswas, M. R. (2015). Regression analysis for prediction of residential energy consumption. *Renewable and Sustainable Energy Reviews, 47*, 332–343.

Gardiner, R., & Hajek, P. (2020). Municipal waste generation, R&D intensity, and economic growth nexus–A case of EU regions. *Waste Management, 114*, 124–135.

Gerami, J., Mozaffari, M. R., Wanke, P. F., & Correa, H. (2021). A novel slacks-based model for efficiency and super-efficiency in DEA-R. *Operational Research*, 1–38.

Giannakitsidou, O., Giannikos, I., & Chondrou, A. (2020). Ranking European countries on the basis of their environmental and circular economy performance: A DEA application in MSW. *Waste Management, 109*, 181–191.

Gökgöz, F. (2010). Measuring the financial efficiencies and performances of Turkish funds. *Acta Oeconomica, 60*(3), 295–320.

Gökgöz, F., & Erkul, E. (2019). Investigating the energy efficiencies of European countries with super efficiency model and super SBM approaches. *Energy Efficiency, 12*(3), 601–618.

Gökgöz, F., & Yalçın, E. (2022). A slack-based DEA analysis for the world cup teams. *Team Performance Management, 28*(1/2), 1–20.

Halkos, G., & Papageorgiou, G. (2016). Spatial environmental efficiency indicators in regional waste generation: A nonparametric approach. *Journal of Environmental Planning and Management, 59*(1), 62–78.

Halkos, G., & Petrou, K. N. (2019). Assessing 28 EU member states' environmental efficiency in national waste generation with DEA. *Journal of Cleaner Production, 208*, 509–521.

Hosseini, S. M., Saifoddin, A., Shirmohammadi, R., & Aslani, A. (2019). Forecasting of CO_2 emissions in Iran based on time series and regression analysis. *Energy Reports, 5*, 619–631.

Jiang, H. (2021). Spatial–temporal differences of industrial land use efficiency and its influencing factors for China's central region: Analyzed by SBM model. *Environmental Technology & Innovation, 22*, 101489.

Jiang, T., Zhang, Y., & Jin, Q. (2021). Sustainability efficiency assessment of listed companies in China: A super-efficiency SBM-DEA model considering undesirable output. *Environmental Science and Pollution Research, 28*(34), 47588–47604.

Joseph, A. M., Snellings, R., Van den Heede, P., Matthys, S., & De Belie, N. (2018). The use of municipal solid waste incineration ash in various building materials: A Belgian point of view. *Materials, 11*(1), 141.

Keskin, B. (2021). An efficiency analysis on social prosperity: OPEC case under network DEA slack-based measure approach. *Energy, 231*, 120832.

Lee, H., Choi, Y., & Seo, H. (2020). Comparative analysis of the R&D investment performance of Korean local governments. *Technological Forecasting and Social Change, 157*, 120073.

Li, J., & Lin, B. (2017). Ecological total-factor energy efficiency of China's heavy and light industries: Which performs better? *Renewable and Sustainable Energy Reviews, 72*, 83–94.

Li, H., & Shi, J. F. (2014). Energy efficiency analysis on Chinese industrial sectors: An improved super-SBM model with undesirable outputs. *Journal of Cleaner Production, 65*, 97–107.

Li, J., Yao, Y., Zuo, J., & Li, J. (2020). Key policies to the development of construction and demolition waste recycling industry in China. *Waste Management, 108*, 137–143.

Long, R., Ouyang, H., & Guo, H. (2020). Super-slack-based measuring data envelopment analysis on the spatial–temporal patterns of logistics ecological efficiency using global Malmquist Index model. *Environmental Technology & Innovation, 18*, 100770.

Malinauskaite, J., Jouhara, H., Czajczyńska, D., Stanchev, P., Katsou, E., Rostkowski, P., & Spencer, N. (2017). Municipal solid waste management and waste-to-energy in the context of a circular economy and energy recycling in Europe. *Energy, 141*, 2013–2044.

Namlis, K. G., & Komilis, D. (2019). Influence of four socioeconomic indices and the impact of economic crisis on solid waste generation in Europe. *Waste Management, 89*, 190–200.

Obersteiner, G., Gollnow, S., & Eriksson, M. (2021). Carbon footprint reduction potential of waste management strategies in tourism. *Environmental Development, 39*, 100617.

Pais-Magalhães, V., Moutinho, V., & Marques, A. C. (2021). Scoring method of eco-efficiency using the DEA approach: Evidence from European waste sectors. *Environment, Development and Sustainability, 23*(7), 9726–9748.

Palafox-Alcantar, P. G., Hunt, D. V. L., & Rogers, C. D. F. (2020). The complementary use of game theory for the circular economy: A review of waste management decision-making methods in civil engineering. *Waste Management, 102,* 598–612.

Ramachandra, T. V., Bharath, H. A., Kulkarni, G., & Han, S. S. (2018). Municipal solid waste: Generation, composition and GHG emissions in Bangalore, India. *Renewable and Sustainable Energy Reviews, 82,* 1122–1136.

Salmenperä, H. (2021). Different pathways to a recycling society–Comparison of the transitions in Austria, Sweden and Finland. *Journal of Cleaner Production, 292,* 125986.

Sarra, A., Mazzocchitti, M., & Nissi, E. (2020). A methodological proposal to determine the optimal levels of inter-municipal cooperation in the organization of solid waste management systems. *Waste Management, 115,* 56–64.

Serge Kubanza, N., & Simatele, M. D. (2020). Sustainable solid waste management in developing countries: A study of institutional strengthening for solid waste management in Johannesburg, South Africa. *Journal of Environmental Planning and Management, 63*(2), 175–188.

Shen, M., Song, B., Zeng, G., Zhang, Y., Huang, W., Wen, X., & Tang, W. (2020). Are biodegradable plastics a promising solution to solve the global plastic pollution? *Environmental Pollution, 263,* 114469.

lo Storto, C. (2021). Effectiveness-efficiency nexus in municipal solid waste management: A non-parametric evidence-based study. *Ecological Indicators, 131,* 108185.

Taboada, G. L., Seruca, I., Sousa, C., & Pereira, Á. (2020). Exploratory data analysis and data envelopment analysis of construction and demolition waste management in the European Economic Area. *Sustainability, 12*(12), 4995.

Tone, K. (2001). A slacks-based measure of efficiency in data envelopment analysis. *European Journal of Operational Research, 130*(3), 498–509.

Tone, K. (2002). A slacks-based measure of super-efficiency in data envelopment analysis. *European Journal of Operational Research, 143,* 32–41.

Tone, K. (2004). Dealing with undesirable outputs in DEA: A slacks-based measure (SBM) approach. Presentation at NAPW III, Toronto, 44–45.

Xiao, C., Wang, Z., Shi, W., Deng, L., Wei, L., Wang, Y., & Peng, S. (2018). Sectoral energy-environmental efficiency and its influencing factors in China: Based on SU-SBM model and panel regression model. *Journal of Cleaner Production, 182,* 545–552.

Yang, Q., Fu, L., Liu, X., & Cheng, M. (2018). Evaluating the efficiency of municipal solid waste management in China. *International Journal of Environmental Research and Public Health, 15*(11), 2448.

Yan, D., Kong, Y., Ye, B., Shi, Y., & Zeng, X. (2019). Spatial variation of energy efficiency based on a Super-Slack-Based Measure: Evidence from 104 resource-based cities. *Journal of Cleaner Production, 240,* 117669.

Yan, D., Ren, X., Kong, Y., Ye, B., & Liao, Z. (2020). The heterogeneous effects of socioeconomic determinants on PM2. 5 concentrations using a two-step panel quantile regression. *Applied Energy, 272,* 115246.

Zambrano-Monserrate, M. A., Ruano, M. A., & Ormeño-Candelario, V. (2021). Determinants of municipal solid waste: A global analysis by countries' income level. *Environmental Science and Pollution Research, 28*(44), 62421–62430.

Zhang, C., Hu, M., Di Maio, F., Sprecher, B., Yang, X., & Tukker, A. (2022). An overview of the waste hierarchy framework for analyzing the circularity in construction and demolition waste management in Europe. *Science of the Total Environment, 803*, 149892.

Zhong, K., Wang, Y., Pei, J., Tang, S., & Han, Z. (2021). Super efficiency SBM-DEA and neural network for performance evaluation. *Information Processing & Management, 58*(6), 102728.

Chapter 6

The Multipronged Approach of Solid Waste Management Toward Zero Waste to Landfill Site: An Indonesia and Thailand Experience

Ariva Sugandi Permana, Sholihin As'ad and Chantamon Potipituk

Abstract

The zero-waste term in municipal solid waste management has been the utopian objective of every waste management authority in the cities in developing countries, even though it comes with different perceptions, which are sometimes misguided. People can produce no waste unless they live with no consumption. The zero-waste term does not mean that we produce no waste, rather we dump no waste at the landfill site. It means we dispose of nothing at a landfill site since the issue of landfill site can be a culprit of waste management, for its reiterating city land demands that generate "headaches" to city authority because of NIMBYism (Not In My Back Yard issue). No one accepts living voluntarily next to a landfill site as it creates more harm than harmless. With zero waste at the landfill site in mind, the waste management authority attempts to deal with the complexity of municipal solid waste management, by reviving each element of the waste management stakeholders to concertedly move on to deal with waste. Individual households and communities, without which waste management will not be successful, were positioned as the main thrust of waste management. A multipronged approach was implemented with all stakeholders, i.e., lawmakers, regulators, waste producers, implementers, and pressure groups, appearing with different functions but a common point: zero waste at the landfill site. A stakeholder with a large capacity, i.e., local government focuses on creating a large project that has a large impact on overall waste management; private sectors may contribute to establishing recycling centers, and waste-to-energy projects. Meanwhile, the individual households,

Pragmatic Engineering and Lifestyle, 113–129

Copyright © 2023 Ariva Sugandi Permana, Sholihin As'ad and Chantamon Potipituk

Published under exclusive licence by Emerald Publishing Limited

doi:10.1108/978-1-80262-997-220231006

which are large in number but have a small capacity, establish community-based activities, i.e., waste banks. This chapter attempts to provide the overall picture of municipal solid waste management in 14 cities in developing countries toward their goal of zero waste at landfill sites.

Keywords: Zero waste at a landfill site; community-based waste management; waste hierarchy; community waste bank; municipal solid waste management; zero waste

1. A General Overview

Certainly, most of us do not like waste since, by definition, waste is something we no longer need, refuse, and trash. However, we cannot avoid producing waste along with the constant increment of consumption as our life is improving and our lifestyle is also changing in an unsustainable direction if we constantly entertain our hedonic needs (Permana et al., 2022). On the other hand, we need to survive living on the only earth. If we continue to live with our current lifestyle, we need theoretically "7 Earths." This is an inconvenient and terrifying truth, and we agree to promote sustainable living. One of the ways to live sustainably is by dealing with the municipal waste generated by the community. A NIMBY is the refusal of society in the presence of landfill sites since they do not want to live next to a landfill site (Johnson & Scicchitano, 2012). This is certainly undesirable from the viewpoint of sustainable waste management. We subscribe to the term sustainable waste management, which is the way we handle solid waste based on the principles of minimum impact on the environment; it could be implemented easily without creating very much social costs; it promotes local economic prosperity. Sustainable waste management must be able to reduce waste production toward manageable waste, and ultimately zero waste state (Seadon, 2010). This chapter was largely based on the first author's research in the Mekong Region through the "Mekong Region Waste Refinery International Partnership Project" funded by the Government of Finland, and a PhD student's research on waste governance in Indonesia under the supervision of the first author.

A multipronged approach to managing municipal waste means that we do all possible efforts that would be able to reduce the quantity of waste disposed to a landfill site. The approaches could range from the waste source, e.g., reducing waste production, to the waste processing domain, from conventional way to nonconventional way such as waste bank including waste valorization. These efforts have been in place for years in cities in developing countries, for example, Jakarta, Bandung, Makassar, Surabaya, Semarang, Surakarta, Medan in Indonesia, and Bangkok, Nakhon Ratchasima, Chiang Mai, Phitsanulok, Pattaya, and Nakhon Sawan in Thailand. The actors could also be varying particularly individual citizens, community-based organizations, nongovernment organizations, private sectors, and government organizations. The discussion in this chapter is based on the experience of undertaking the multipronged approaches in Indonesia and Thailand. The reasons for selecting these two countries are because the authors have substantial experience in these two countries. The discussion would not thematically be distinguished based on these two countries but it would rather complement one another. Recorded experiences in other countries in Southeast Asia would also be discussed if possible. The municipal waste management in this regard is solid waste management excluding hazardous waste. Aggregate impacts on regional administration or political entities such as the Southeast Asia region could not be reflected here as well, as these two countries could not represent SE-Asia.

Based on ten-year research on waste management in Cambodia, Indonesia, Lao PDR, Thailand, and Vietnam, the local government in developing countries of Southeast Asia failed to accomplish sustainable waste management because of some factors. Our research shows that this is because (1) they have simplified the process of waste management by only collecting, transporting, and disposing of the waste, (2) while the relevant waste regulations are sufficiently available, the implementation is inadequate, (3) a conventional and persistent "developing country issue" exists, which is insufficient budget allocation to carry out sufficient municipal waste management, (4) despite ubiquitous sustainable waste management knowledge, insufficient equipment and workforce are persistent, and (5) lack of willingness to promote the current state of waste management to a higher level. This is because of the unwillingness to move from the comfort zone to explore new things and new challenges, which require significant sacrifice.

Based on the authors' research and observation during the period of 2005–2010 in Cambodia, Indonesia, Lao PDR, the Philippines, Thailand, and Vietnam, if we connect the lifestyle, which is reflected in the level of consumption and waste production, and environmental integrity, the path, present status, and goal of the waste management could be found here, as schematically displayed in Fig. 1. In this figure, sustainability is measured by three characteristics of the environmental aspect (low waste production and pollution), the social aspect (social welfare), and the economic aspect (prosperity).

Southeast Asian countries, as represented by Indonesia, Lao PDR, Thailand, and Vietnam, are generally in low waste production status because they are yet to be industrialized, and therefore the waste production and consumption level are considerably low. They are also in a low-sustainability state because their prosperity and social welfare are far below the socioeconomically sustainable state. They can keep on the path ① toward the ultimate goal, which is zero waste. However, it is almost impossible as it requires the lowest consumption with zero-waste production. The rational path is through the path ②, in which the capacity of municipal waste management is upgraded to moderate and later to high capability waste management. On the other hand, the developed countries confront only reducing waste production and enjoy an easier path toward zero waste, while keeping economic development cherished. This is because developed countries have the power of technology, strong financial capability, and leading human resources, which unarguably can "buy" (accomplish) many things including sustainable development. As exhibited in Fig. 1 or Fig. 2, developed countries can reduce their production of waste through science and technology, and at the same time, consistently advance their economies.

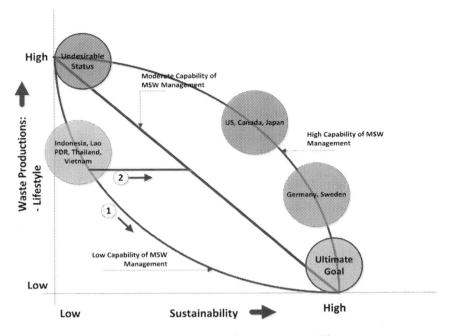

Fig. 1.   Lifestyle vis-à-vis Sustainability.

Table 1 shows the correlation between waste production, per capita GDP, and other associated variables. The table shows that there is no consistent pattern correlation between the GDP and waste production for the mixed aggregate developed countries and developing countries. However, if we separated the developed countries and the developing countries, the correlation seems more consistent for both despite outlier countries, which are Canada (for the developed countries group), and Lao PDR (for the developing countries group).

Similar to Fig. 1, a more concrete correlation can also be exhibited through the relationship between per capita GDP and per capita waste production, adding three countries with different levels of per capita GDP, as shown in Fig. 2.

Fig. 2 shows a similar pattern to Fig. 1, in which Canada, the United States, Japan, Germany, and Sweden are in the same group of high-income countries keeping the waste per capita production low. Thus, the shaded region (lower-right part) of Fig. 2 would be the ideal condition for sustainable development, in which economic development goes hand-in-hand with the advancement of environmental aspects as reflected in the low per capita waste production. Theoretically, the group of Malawi, Vietnam, Indonesia, Thailand, Guyana, and Oman with lower waste production has only linear one-dimensional effort on the way to accomplishing a sustainable development state, providing that they can keep waste production low during their economic advancement process. On the other

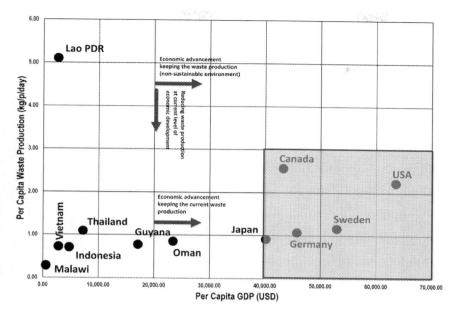

Fig. 2.   Per Capita Waste Production and GDP of Selected
Countries.

hand, Lao PDR must go for two-dimensional efforts, i.e., lowering waste production and advancing the economy to achieve a sustainable development state. However, many city governments failed during this process. One of the indications of the failure is that, in Indonesia, the demand for landfill sites has been more frequent from 20-year intervals in the 1990s to only 8-year intervals in 2015 (Permana & Towolioe, 2019).

The failure of many city governments in Indonesia in managing their municipal waste was due to the above five points that have been explained earlier, particularly the simplified municipal waste management of collecting, transporting, and disposing thinking (Permana & Towolioe, 2019). Their mindset of the accomplishment of municipal waste management was only measured from the "cleanliness of the city," which is purely framed by a collecting-transporting-disposing mentality. The issue of repeating the needs of a landfill site in 5–10 years intervals amid limited land for the site and NIMBYism has never been coming into their mind. As a result, a pseudo-complacency of municipal waste management has been embedded. For example, in Indonesia, The *Adipura* (an award for Cleanest City) was based merely on the cleanliness of the city, no waste remains uncollected in the city, and the city looks nice, and the city will get this award. It was not based rigorously on broader context, i.e., sustainable city than merely city cleanliness.

Table 1. Waste Production and GDP.

Country	Per Capita GDP (USD)	Annual Waste Production (Million Ton/ year)	Population (Million)	Per Capita Waste Production (Ton/Year)	Per Capita Waste Production (kg/day)
USA	63,543.58	268.00	329.50	0.81	2.23
Sweden	52,925.71	4.40	10.35	0.43	1.16
Germany	45,723.64	32.50	83.20	0.39	1.07
Canada	43,241.62	35.60	38.01	0.94	2.57
Japan	40,113.06	41.70	125.80	0.33	0.91
Oman	23,416.00	1.58	5.10	0.31	0.85
Guyana	17,108.00	0.22	0.79	0.28	0.77
Thailand	7,189.04	27.80	69.80	0.40	1.09
Indonesia	4,722.00	70.00	273.50	0.26	0.70
Vietnam	2,785.00	25.50	97.40	0.26	0.72
Lao PDR	2,630.00	13.60	7.30	1.86	5.10
Malawi	545.00	1.96	19.13	0.10	0.28

Source: UN-Desa, UN-Habitat, World Bank, IMF, and government agencies.

2. Present Development of Waste Management Toward Zero Waste

Mason et al. (2003), Erkelens (2003), and Lehmann (2010) assert that many municipal governments are aware and willing to attain zero waste to the landfill site in their solid waste management program since zero waste gained popularity and was a keyword in urban development. The term zero waste was introduced in the 1990s in Europe. During this period, the advocacies were focused on environmental sustainability with sustainable consumption and production (Geels et al., 2015; Lorek & Spangenberg, 2014). The term "zero waste" does not eliminate the waste from the source, but rather minimizes waste production by minimizing consumption along with 3Rs and waste recovery. Zero waste was inspired by the fact that land resources are limited particularly in urban areas to cater to the growing need for landfill sites (Gallo, 2019; Sebastien, 2017; Simsek et al., 2014). Citizens attempted to explore minimizing waste disposal to the landfill site and therefore eliminating the persistent needs of landfill sites along with the possible social and environmental effects. Despite its insignificance, the landfill site might generate income for waste scavengers and recyclables collectors. However, the negative effects generated by the activity are much higher than the benefits gained by the scavengers (Asim et al., 2012; Besiou et al., 2012; Ferronato & Torretta, 2019; Periathamby et al., 2009).

Fig. 3 shows the present situation of waste flow (from source to landfill) in the Mekong Region countries, i.e. Thailand, Lao, Cambodia, and Vietnam, less the accomplishment of zero waste to the landfill site. The willingness of the local governments in this region to attain zero waste at the landfill site was high with the assistantships provided by CIDA Canada, SEI Sweden, and the Government of Finland. The pilot projects of zero waste were more than enough, but when it comes to the scale-up, replication, and real implementation (not a pilot project), the curse of developing nations of "good in planning but worse in implementing" hampered the program. As reflected in the figure, a high percentage (70–80%) of biodegradable waste in the region is present in the Southeast Asia region. The pilot projects in Bangkok suggested developing composting facilities, and biogas digesters at community and individual levels. Bangkok Metropolitan Government operates a large composting facility with a capacity of 50–75 tons per month. However, the demand for composting products was not that high, which determines the production cycle of the compost to absorb the increasing quantity of biodegradable waste. In the Bangkok Metropolitan itself, the biodegradable waste production was about 1,000–1,500 tons per day in 2012 (Personal communication with Bangkok Metropolitan Administration, 2012). The compost production of 50–75 tons per month in 2012 was too small in comparison to biodegradable waste production in Bangkok. Therefore, the production capacity must be increased, and the Citizens of Bangkok could be involved. The demand

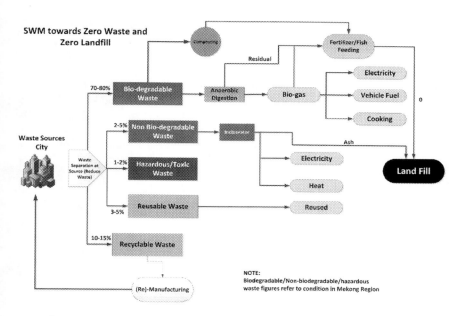

Fig. 3.   From Source to Landfill Toward Zero Waste in Mekong Region.

for compost products should also be increased significantly. The compost product is commonly used for the replacement of chemical fertilizer. However, most people prefer to use chemical fertilizer over compost for the reason of agricultural production, even though the cost of chemical fertilizer is much higher than the composting product. At this point, composting cannot economically and sustainably resolve the biodegradable waste to accomplish zero waste to the landfill site, unless the economic consideration is set aside, which means the existing composting facilities must be subsidized by the government. Innovation in compost must be continuously created to make compost better and more economic than chemical fertilizer.

The biodegradable waste can also produce biogas for generating electricity, vehicle fuel, and cooking gas, through the construction of anaerobic digesters. Based on project observation in Pitsanulok and Nakhon Ratchasima (Thailand), this process does not eliminate the waste since the residual of the anaerobic digestor is still large, i.e., 30–40%. The residuals can go to either open land to reutilize the land or to the incinerator, to reduce further the residuals, and finally, it goes to landfill sites. Observation at the Mekong Region Waste Refinery Project reveals that biogas production was more economically viable than compost in Thailand, for a reason that the number of consumers of biogas much more than the compost product. The processing of biodegradable waste through composting and biogas production is a potential approach to zero waste. Solving the problem with biodegradable waste in the region means unraveling most of the solid waste management problems since biodegradable waste is the predominant waste.

Some undergoing waste-to-energy project in Indonesia is, for example, the Putri Cempo Project in the City of Surakarta, Indonesia, with an electric generating capacity of 10 MW. This waste-to-energy project requires 545 tons of flammable waste a day, while the flammable waste products of the city are 260 metric tons a day (Personal communication with the Local Government of Surakarta City in 2021). This shows that this project resolved most of the waste problem in the city. The city will no longer need the landfill site to dispose of the solid waste, and even contribute to solving part of the solid waste problem in the neighboring city as the deficit of the flammable waste is around 285 metric tons per day. Similar waste-to-energy projects are currently undergoing in Jakarta, Surabaya, and Makassar.

Two waste-to-energy plants that produce 35 MW of electricity are currently in operation in Nongkhaem and Onnut Districts in Bangkok. They require around 2×1000 metric tons of waste per day for firing the plants. These plants theoretically reduce the quantity of waste at the landfill site by as much as $2000 - (0.04 \times 2000) = 1920$ metric tons [NOTE: 0.04 is approximate residuals that finally go to landfill site]. This is a hypothetical illustration that if the approximate daily waste production in Bangkok was 10,000 metric tons, these two plants alone have contributed significantly (about $1920/10,000 = 19.2\%$) to waste disposal at the landfill site. A waste-to-energy plant produces typically the remaining 0.5–8.0% of the ashes or residues. In the waste hierarchy, the waste-to-energy or waste recovery program is the second of the least-preferred way of reducing waste. It is easier to implement in comparison to waste prevention, the most preferred, and the hardest to carry out because it affects the overall people's lifestyle.

Waste recycling activities in Indonesia and Thailand were almost similar. However, Thailand is at a little bit advance stage as many recycling companies are in operation to recycle the waste. The recycling activities are carried out through two bodies, which are waste banks and waste recycling itself. The waste banks were operated mostly by communities, while the recycling companies were private sectors (Permana et al., 2015; Permana & Towolioe, 2019).

3. Society-Based Waste Management Toward Zero Waste

One school of thought argues that bottom-up planning has a higher probability of success if implemented compared to top-down planning (Long & Franklin, 2004; Sabatier, 1986). A key reason is that in bottom-up planning, everyone is encouraged to participate and many people are involved, leading to a higher sense of responsibility for what has been agreed and decided upon. In contrast, top-down planning requires only a few people to involve, and if implemented only a few people have a sense of responsibility. It is easier to implement and manage. The larger number of people meanwhile left uninvolved, and could probably deny it. The only advantage of top-down planning is easy and quick.

Similar logic can also be applied to municipal waste management. Citizens are potential actors for the success of waste management, for their large number. The greater number of involvements in municipal waste management, the chance for success is higher. Permana et al. (2015) asserted that successful municipal waste management can be measured quantitatively from the percentage of collected municipal waste, and the percentage of citizens' perceptions of the cleanliness of their city. Another sustainability indicator is if the ideal percentage of the waste hierarchy is accomplished or on the right path to the ideal waste hierarchy. The ideal waste hierarchy from the social and environmental sustainability viewpoint and the existing waste hierarchy in most cities in Indonesia and Thailand is schematically depicted in Fig. 4. The ideal waste hierarchy shows the approximate percentage of each element of waste management. From a sustainability viewpoint, waste prevention is the most preferred and waste disposal is the least preferred element of waste management.

The discrepancy between the ideal and existing situations in some cities in Indonesia, for instance, Makassar, Bandung, Surabaya, Medan, and many more as represented by Fig. 4 is sufficient to say that the waste management carried out by the local government has failed to accomplish the environmentally sustainable waste management.

The element of municipal (solid) waste management in this multipronged approach can be extracted from Figs. 3 and 4. The elements, in which individuals and/or society as the prime mover, are discussed in the following subsections. Five elements of waste management have been done and continue to improve in Indonesia and Thailand. These five elements include waste valorization, community waste recycling, waste bank, waste composting, and waste digester.

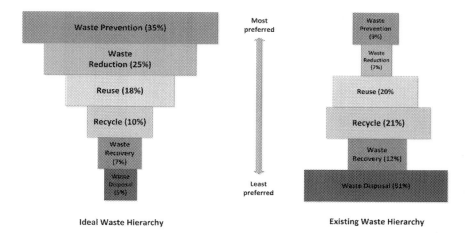

Fig. 4.   The Waste Hierarchy: Ideal vis-à-vis Existing Status of Thailand.

3.1 Waste Valorization

Waste valorization means any process that would be able to convert the waste into something useful and valuable or contains a value, with or without changing the total physical appearance of waste. For example, waste to best, which is changing the discarded car tire to be a useful garden chair (Fig. 4). The waste valorization is also in line with the circular economy that attempts to keep the waste containing economic values as long as possible, and therefore it will disturb the environment at the shortest possible time during its lifecycle. By this definition, many waste activities are directly or indirectly connected with waste valorization, for example, waste to energy, waste reuse, and waste composting. Waste valorization is possible with creative and out-of-the-box thinking.

Bali and Yogyakarta, the two most tourist attractions of Indonesia because of cultural competitiveness, have numerous creative thinkers converting waste into something valuable through a handy-crafting process. The values after undergoing valorization can be a 100 folds or more from the original value of the waste. For example, the chair exhibited in Fig. 5 has infinite folds of the original value. The original value of the tire is zero or almost zero, while the selling price of the end product is about USD 75. Therefore, this valorization has increased, theoretically, the value of waste to infinity (USD75/0 = ∞). This selling price of USD 75 is the real value enjoyed by handcrafters with practically zero investment. The valorization can only be done with creative and out-of-the-box thinking, and not all people can do the waste valorization.

Fig. 5.   A Waste Valorization in Circular Economy. *Source:* Anonym.

3.2 Community's Waste Recycling

In Indonesia and Thailand, waste recycling is commonly carried out by the large factory owned by private companies. An individual, in most cases, cannot do waste recycling since it requires knowledge, tools, equipment, and money. Individuals with these four requirements can do waste recycling. A community with a cooperative that will be able to procure the basic requirements can establish and run a waste recycling facility. Metals and plastic recycling can be done at the community level with the possession of these four basic requirements.

Some observations in small-scale industries run by individuals and groups of individuals in Indonesia were able to produce water pumps, spare parts for vehicles, bicycles, motorcycles, etc. With the guidance provided by the authority, their products are standardized with cheaper than the original spare parts. Even, with the quality control offered by original manufacturers, the local products can compete with other similar products. However, recycling, particularly metal recycling has a very low contribution to waste disposal at landfill sites, since all metals are valuables, and without being expressed, the metals are most certainly going to recycling centers.

3.3 Waste Bank

A waste bank is a banking-like system, in which the object of the transaction is waste. The member or customers of the waste bank can be anyone in the community, waste pickers in particular. A member will have a waste bank book provided by the waste bank. The CEO of the Waste Bank is an individual or waste collector. A valuable waste, particularly recyclables are valued in monetary term, for example, one a-1 liter plastic bottle is worth 8 to 16 Cent, and so on. The

Waste Bank determines the price of the recyclables. Saving in a Waste Bank could be done by a member of the Bank by bringing the recyclables to a Waste Bank. The Bank carries out a valuation on the recyclables brought by the member and records the saving in the bank's book. After some time, where the saving accumulated large enough and has sufficient value, the member can withdraw an amount of money from the Waste Bank, or just simply leave it for further accumulation. The recyclables accumulated by Waste Bank are sold to the recycling center at the price higher than that bought from the member. The price difference is accumulated by the Waste Bank as a Bank's profits and used for the operation of the Bank. Through this process, the collection of recyclables by the recycling center can be more speedy and more organized.

In Makassar City Indonesia, there were 78 active and in-operational waste banks with a total active member of 30,000. On average, one member was able to save about IDR 250,000 (USD 16) monthly. The recorded highest saving of a member was about IDR 5,300,000 (USD 350) per month. The waste bank involved a financial transaction of around IDR USD 400,000 in a year. This is a small amount for even a normal village bank, but it is a significant number of transactions for a small enterprise in a city. In short, the waste bank has been able to provide additional income to urban poor families, particularly waste pickers.

The way waste banks contribute to the reduction of waste disposed at landfill sites can be estimated from the baseline survey before the establishment of the waste bank, and the additional quantity of the recyclables sent to the recycling

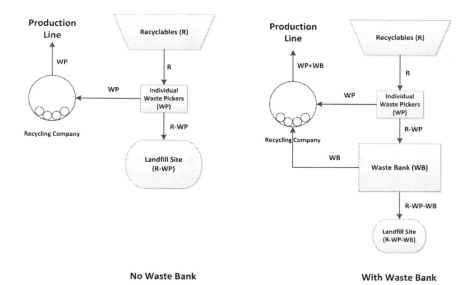

Fig. 6.   Contribution of Waste Bank to the Zero Waste at Landfill Site.

center (Fig. 6). In short, the contribution of the waste bank to zero waste at the landfill site is expressed by the following equation:

$$\text{Contribution} = (R - \text{WP} - \text{WB}) - (R - \text{WP}) = -\text{WB}$$

It is negative because the Waste Bank diverts the recyclables from the landfill site to the recycling center and goes back to the production line (useful product). The size of the landfill site, in which the waste bank is part of the municipal solid waste management (MWSM) system, is smaller than where the waste bank does not exist in the MSWM system. All the transactions in the waste bank are using recyclables, particularly various types of plastic bottles, i.e., Polyethylene Terephthalate (PET), Polypropylene (PP), High-Density Polyethylene (HDPE), and Low-Density Polyethylene (LDPE). The waste bank in Makassar City has been able to divert part of the recyclables from the landfill site to the recycling center of around 480 metric tons per year out of total waste production in Makassar City of 900 metric tons per day. The contribution is insignificant, but still, the contribution to zero waste at the landfill site is commendable.

In Bangkok, a similar waste bank with a similar system, i.e., run and managed totally by the community have been in place for a few decades now. Unlike in Makassar City, in Bangkok, the waste bank received different kinds of recyclables, not only plastic bottles but also metal scraps. The Banks collaborate with Primary Schools and Buddhist Temples to promote the waste bank and increase the intensity of recyclables collection. By this arrangement, the recyclables collected could be expedited, as most primary schools were involved. There was no record found on the quantity of financial involvement in the community-based waste bank in Bangkok. However, looking at the population size and the intensity of involvement of the primary schools and temples in Bangkok, there were 10–15 times larger than the volume of the waste bank in Makassar City Indonesia.

3.4 Waste Composting

Individual households in Makassar City did waste composting using a so-called Magic Takakura Composter. A 5–10 liter basket with holes in its wall surroundings. The process is very simple, just place the food waste mixed with soil, and maintain it for a few days, the waste magically turned into a compost product. This Takakura composter could also be replaced by a similar plastic basket or container. Due to its simplicity, and government subsidy on the provision of Takakura Composter, the composting activities in the City were very popular. We recorded about 20,000 citizens involved in the composting activities. Some citizens use the compost product for their purpose or sell it directly to other citizens. However, a problem arises when the compost product was not absorbed by the market while also saturation of the demand does exist in the locality.

Our study shows that there is currently no factory-size waste composting facility, which is extensively producing compost products, found in Indonesia. As agricultural activities are still a significant sector in Indonesia, and therefore the composting product has a good prospect. However, most farmers preferred chemical fertilizer to biodegradable fertilizer like compost. From the product viewpoint, the use of compost does not look quite promising in comparison to chemical fertilizer. These are some solid reasons the farmers do not use compost for promoting their products. The challenge here is how to increase the productivity of farming without expensive chemical fertilizers by using compost. If this simple question can be perfectly solved, the solution brings multiple positive externalities to the environment, i.e., increasing farming products without much disturbance to the environment, and also reducing waste disposed at the landfill site.

The waste composting, which is carried out individually or by individual households in Bangkok, was not that strong in comparison to Indonesia in general. However, Bangkok has a factory-scale composting facility as mentioned earlier, which is unavailable in Indonesia. However, in rural areas, it seems that composting activities in connection with solid waste management is existing in a not so rapid pace. Some cities in Central and East Java have factory-scale composting facilities; however, the production size is not as large as the Bangkok one, and few composting facilities have already stopped their production for the same reason: lack of demand. The contribution of waste composting to zero waste at a landfill site can be approximated from the total quantity of organic waste, i.e., kitchen waste and other composting materials converted to compost products, i.e., useable products.

3.5 Waste Digesters

In typical cities in Indonesia and Thailand, organic waste is the largest and most prevailing quantity of waste. It contains 70–80% organic waste or wet waste or biodegradable waste (Permana et al., 2015). Resolving this type of waste means resolving 70–80% of the waste problems. One of the solutions to this organic waste is by converting the waste into biogas. Biogas can generate electricity, and gas for cooking and heating. Our past project promoted portable biogas digesters to be used by individual households along with one or two community biogas digesters. A biogas digester with a capacity of equivalent generated electricity of 200 kW costs around USD 600 in Thailand. This is suitable for the use of the individual household. For the community digester, the cost might be up to USD 5000.

One of the disadvantages of biogas production, from the zero-waste viewpoint, is that it might produce a high amount of residual waste such as ash and sewage. The unburnt biogas can also produce methane, which is a greenhouse gas, and this issue requires specific attention to avoid producing additional environmental impact. The biogas digester could be designed to serve the size of individual households and communities. Our past project in Thailand was focused on the biogas digester of the individual household. However, at the community level, a digester with a maximum capacity of 100 kg of waste input per day was also constructed. The waste inputs were mostly kitchen waste supplied by individual households and wet waste from the market's waste disposal, i.e., grocery waste. A rough estimate of 100 kg of biodegradable waste may produce around 231 MJ of energy per day for a period maximum of 45 days. However, some input materials might still produce biogas for a period longer than 45 days. Table 2 shows the biogas production and period of production.

Another way to produce energy from biodegradable waste is to produce incendiary materials to yield heat by burning the material in the incinerator-cum-boiler and steam turbine to produce electricity. This process needs a large investment for the system, in comparison to constructing biodigester to produce biogas. The approximate electric production through a waste-to-energy program for various types of waste is shown in Table 3. The advantage of this process is that almost all wastes, not only biodegradable, are combustible. Therefore, this program can be placed at the most downstream part of the municipal waste management system toward zero waste at a landfill site, as exhibited in Fig. 3.

4. The Way Forward

A zero-waste program is not impossible to accomplish with strong willingness from the stakeholders, particularly local government, as the regulator and policymaker, as our observation suggested. Municipal solid waste management stakeholder is commonly consisting of local government, citizens, private sectors, nongovernment organizations, and community-based organizations. The role of the community is central to municipal solid waste management as we observed in Makassar City, Indonesia, as well as some secondary cities in Thailand. The role of local government as regulator and implementer of municipal solid waste management is also crucial, without which the management will not be running well, particularly when the citizens or communities are having very low awareness of waste management. For instance, our observation in Makassar City shows that waste bank runs under high-awareness citizens and it was almost independent of the participation of local government. However, this situation may not reflect the overall picture of the community in Makassar. In Bangkok, active primary schools and communities, to some extent have helped the high awareness of

Table 2. Biogas Production and the Period of Production.

Raw Material	Biogas Yield per kg of Fermented Materials (in m³)	Percentage of Biogas Production in Days		
		0–15	15–45	45–75
Cow dung	0.12	11.0	33.8	20.9
Pig manure	0.22	19.6	31.8	25.5
Human waste	0.31	45.0	22.0	27.3
Water Hyacinth	0.16	83.0	17.0	0.0
Rice straw	0.23	9.0	50.0	16.0

Source: Based on Szyba and Mikulik (2022), Baud et al. (2004).

Table 3. The Approximate Potential Production of Electricity for Various Types of Waste per 1,000 kg Biodegradable Waste.

Raw Material	Biogas Yield (m³)	Electricity Yield (kWh)	Heat Produce (MJ)
Grocery store waste	75.0	170.0	900
Trimmed grass	175.0	342.0	1,930.0
Maize silage	185.0	331.0	1,871.0
Cattle manure, i.e., cow dung	45.0	88.0	497.0
Chicken manure	80.0	156.0	882.0
Pig slurry	45.0	95.0	536.0

Source: Based on Szyba and Mikulik (2022), Baud et al. (2004).

the communities surrounding. The findings show that the waste management activities done by individual households and communities are the prerequisite for the next level of municipal solid waste management and need to be strengthened with continuous encouragement, leadership exemplary, and financial stimulation toward zero waste at the landfill site.

Acknowledgments

This manuscript was based on a research carried out by the first author under the Project Grant funded by the Government of Finland of the "Mekong Region Waste Refinery International Partnership Project," Grant No. 3-R-091, and a PhD student's research under the supervision of the first author. The authors acknowledge the support of the Government of Finland, and Local Government of Makassar City, Indonesia.

References

Asim, M., Batool, S. A., & Chaudhry, M. N. (2012). Scavengers and their role in the recycling of the waste in Southwestern Lahore. *Resources, Conservation and Recycling, 58*, 152–162.

Baud, I., Post, J., & Furedy, C. (2004). *Solid waste management and recycling: Actors, partnerships, and policies and Hyderabad, India, and Nairobi, Kenya.* Springer Science & Business Media.

Besiou, M., Georgiadis, P., & Van Wassenhove, L. N. (2012). Official recycling and scavengers: Symbiotic or conflicting? *European Journal of Operational Research, 218*(2), 563–576.

Erkelens, P. A. (2003). Re-use of building components (towards zero waste in renovation). In *Proceedings of the 11th Rinker International Conference, CIB Deconstruction and Materials Reuse* (Vol. 287, pp. 7–10). CIB Publication.

Ferronato, N., & Torretta, V. (2019). Waste mismanagement in developing countries: A review of global issues. *International Journal of Environmental Research and Public Health, 16*(6), 1060.

Gallo, M. (2019). An optimisation model to consider the NIMBY syndrome within the landfill siting problem. *Sustainability, 11*(14), 3904.

Geels, F. W., McMeekin, A., Mylan, J., & Southerton, D. (2015). A critical appraisal of Sustainable Consumption and Production research: The reformist, revolutionary and reconfiguration positions. *Global Environmental Change, 34*, 1–12.

Johnson, R. J., & Scicchitano, M. J. (2012). Don't call me NIMBY: Public attitudes toward solid waste facilities. *Environment and Behavior, 44*(3), 410–426.

Lehmann, S. (2010). Resource recovery and materials flow in the city: Zero waste and sustainable consumption as paradigms in urban development. *Sustainable Development Law & Policy, 11*, 28.

Long, E., & Franklin, A. L. (2004). The paradox of implementing the government performance and results act top-down direction for bottom-up implementation. *Public Administration Review, 64*(3), 309–319.

Lorek, S., & Spangenberg, J. H. (2014). Sustainable consumption within a sustainable economy–beyond green growth and green economies. *Journal of Cleaner Production, 63*, 33–44.

Mason, I. G., Brooking, A. K., Oberender, A., Harford, J. M., & Horsley, P. G. (2003). Implementation of a zero-waste program at a university campus. *Resources, Conservation and Recycling, 38*(4), 257–269.

Periathamby, A., Hamid, F. S., & Khidzir, K. (2009). Evolution of solid waste management in Malaysia: Impacts and implications of the solid waste bill, 2007. *Journal of Material Cycles and Waste Management, 11*(2), 96–103.

Permana, A. S., Bandhari, B. S., & Ovhal, N. A. (2022, June). The engineering of zero waste: Between sustainability and waste production. *Innovative Engineering and Sustainability Journal, 1*(1), 38–49.

Permana, A. S., & Towolioe, S. (2019). Why does many local governments fail in managing municipal solid waste? In *Contemporary urban life and development.* Penerbit UNS Press.

Permana, A. S., Towolioe, S., Aziz, N. A., & Ho, C. S. (2015). Sustainable solid waste management practices and perceived cleanliness in a low-income city. *Habitat International, 49*, 197–205. http://dx.doi.org/10.1016/j.habitatint.2015.05.028. Impact Factor: 4.3 (2019).

Sabatier, P. A. (1986). Top-down and bottom-up approaches to implementation research: A critical analysis and suggested synthesis. *Journal of Public Policy, 6*(1), 21–48.

Seadon, J. K. (2010). Sustainable waste management systems. *Journal of Cleaner Production, 18*(16–17), 1639–1651.

Sebastien, L. (2017). From NIMBY to enlightened resistance: A framework proposal to decrypt land-use disputes based on a landfill opposition case in France. *Local Environment, 22*(4), 461–477.

Simsek, C., Elci, A., Gunduz, O., & Taskin, N. (2014). An improved landfill site screening procedure under NIMBY syndrome constraints. *Landscape and Urban Planning, 132*, 1–15.

Szyba, M., & Mikulik, J. (2022). Energy production from biodegradable waste as an example of the circular economy. *Energies, 15*, 1269.

Chapter 7

Conservation; Waste Reduction/Zero Waste

Shima Yazdani and Esmail Lakzian

Abstract

Currently, waste is regarded as a symptom of inefficiency. The generation of waste is a human activity, not a natural one. Currently, landfilling and incinerating wastes are common waste management techniques; but the use of these methods, in addition to wasting raw materials, causes damage to the environment, water, soil, and air. In the new concept of "Zero Waste" (ZW), waste is considered a valuable resource. A vital component of the methodology includes creating and managing items and procedures that limit the waste volume and toxicity and preserve and recover all resources rather than burning or burying them. With ZW, the end of one product becomes the beginning of another, unlike a linear system where waste is generated from product consumption. A scientific treatment technique, resource recovery, and reverse logistics may enable the waste from one product to become raw material for another, regardless of whether it is municipal, industrial, agricultural, biomedical, construction, or demolition. This chapter discusses the concept of zero landfills and zero waste and related initiatives and ideas; it also looks at potential obstacles to put the ZW concept into reality. Several methods are presented to investigate and evaluate efficient resource utilization for maximum recycling efficiency, economic improvement through resource minimization, and mandatory refuse collection. One of the most practical and used approaches is the Life Cycle Assessment (LCA) approach, which is based on green engineering and the cradle-to-cradle principle; the LCA technique is used in most current research, allowing for a complete investigation of possible environmental repercussions. This approach considers the entire life cycle of a product, including the origin of raw materials, manufacturing, transportation, usage, and final disposal, or recycling. Using a life cycle perspective, all stakeholders (product designers, service providers, political

Pragmatic Engineering and Lifestyle, 131–152

Copyright © 2023 Shima Yazdani and Esmail Lakzian

Published under exclusive licence by Emerald Publishing Limited

doi:10.1108/978-1-80262-997-220231007

and legislative agencies, and consumers) may make environmentally sound and long-term decisions.

Keywords: Municipal solid waste; landfill; waste management; sustainability; zero waste; life cycle assessment (LCA)

Nomenclature

APC—Air Pollution Control
GDP—Gross Domestic Product
GSCM—Green Supply Chain Management
HDB—Housing Development Board
LCA—Life Cycle Assessment
MSWM—Municipal Solid Waste Management
SDG—Sustainable Development Goals
UN—United Nations
WtG—Waste-to-Energy
ZL—Zero Landfill
ZW—Zero Waste

1. Introduction

Municipal Solid Waste Management (MSWM) is a global problem affecting human health, the environment, society, and the economy. Waste management often aims to protect the environment and human health while conserving resources. Nonetheless, waste management issues are emphasized to varying degrees depending on the region's way of life. Ongoing development goals (Brunner & Fellner, 2016) include increasing waste collection service coverage and reducing illegal or uncontrolled disposal (transition to sanitary landfilling). In contrast to impoverished or transitional nations, where spending more than 10 euros per person on waste collection, treatment, and disposal is unrealistic, the emphasis is on reducing output and enhancing prevention and resource recovery (Brunner & Fellner, 2016). Heat and mechanical methods are often used. Because of unplanned urbanization, fast population increase, and significant health issues caused by inadequate health infrastructure, emerging countries have particular MSWM difficulties (Khatib, 2011). However, due to governments' limited power to control trash, MSWM is often insufficient and ineffective because of customers who refuse to pay for services, inadequate or unlawful disposal, a lack of regulatory oversight, and so on (Guerrero et al., 2013; Zaman, 2014). MSWM has been developed and enhanced in several sectors through several programs, some of which are based on indicators.

It is critically important to minimize global energy consumption and develop renewable energy sources in light of the scarcity of energy resources. According to Guerrero et al. (2013), a successful system should recognize the links among the environment, society, and economy to promote the effective management of municipal solid waste (MSW) across the system. Addressing environmental goals based on the percentage of treated MSW and population, according to Yu and Solvang (Yu & Solvang, 2017), enhances the possibility of lowering MSW using sustainable and ecologically acceptable techniques. Meanwhile, incorrect or abandoned MSWM has serious detrimental effects on the environment, biodiversity, social welfare, and the economy (Sisto et al., 2017). Poor MSWM performance endangers the environment and human health, particularly when waste is disposed of incorrectly near inhabited areas, water sources, and sewage systems (Coban et al., 2018). MSW output has grown due to population growth, economic development, urbanization, and industrialization (Hamid et al., 2020). MSW is the term used to describe the everyday solid waste created by homes, businesses, and organizations. It includes waste from supermarkets, hotels, offices, retail outlets, and public services such as hospitals, markets, yards, and amusement parks. According to Su et al. (2007), municipal authorities are finding it increasingly difficult to guarantee efficient and sustainable MSWM as the repercussions of MSW become more complicated. Municipal solid garbage is continually inadequate and inefficient due to the government's local waste handling capacity. Human health, environmental preservation, economic progress, and social welfare satisfaction must all be achieved via MSWM solutions (Soltani et al., 2015).

Through a project supported by the European Council, the ZW idea was created (Curran & Williams, 2012). Additionally, it is recommended to challenge conventional wisdom and reduce industrial waste. Slag will be used to maximize its material and energy potential, leading to "zero resource waste." The future vision envisions a world where virtually all input materials and the majority of input energy can be converted into valuable products and recoverable energy (Yang et al., 2022).

The United Nations (UN) established the Sustainable Development Goals (SDGs) in 2015 (Flachenecker & Rentschler, 2018) to achieve a dramatic change in economic structure toward a more sustainable economy by 2030. The emphasis on resource efficiency, a crucial pillar for achieving the long-term strategic goal of sustainable production, is a vital element of these SDGs (İncekara, 2022). The resolution of critical issues, such as resource depletion and waste management, as well as the promotion of sustainable development and innovation toward a circular economy depend on businesses enacting policies that promote more responsible and efficient resource usage (Aristei & Gallo, 2021). Consequently, a resource-efficient circular economy may result in more sanitary industrial processes, which may have various direct and indirect consequences on societal welfare (Özbuğday et al., 2020). From 2018 to 2050, it is anticipated that the proportion of energy used by energy-intensive industries will remain practically constant at 50%. During the same period, the energy consumption rate by non-energy-intensive firms may increase from 35% to 38% (İncekara, 2022). The industrial sector uses around 25% of the world's resources. Improving the "sustainability" of this industrial sector by reducing its energy and resource use is a top objective.

Resource efficiency strategies stress the distribution of goods that can be recycled after a reasonable period, the use of environmentally friendly materials, and raw material packaging. They include initiatives to safeguard resources, reduce waste generation, increase recycling efforts, and reduce pollution. Resource efficiency is maximized when a competitive advantage can be achieved with minimal usage of raw resources or impact on the environment while not losing quality or consistency (Campitelli et al., 2019). Procedures for resource efficiency, such as waste management and process reengineering, may be distinguished by their look and function. Process reengineering approaches strive to reorganize corporate operations to uncover improvement opportunities and develop solutions for resource optimization, such as energy and water savings (Bodas-Freitas & Corrocher, 2019). Meanwhile, waste management plans successfully encompass the actions and activities required to manage rubbish from generation to disposal (Aristei & Gallo, 2021). These strategies reduce waste by modifying processes, reusing, and recycling.

It implies that product life cycle planning should not be confined to the point of disposal but should also include the reintroduction of solid waste into new manufacturing processes (Pietzsch et al., 2017) or reuse in different ways (Smol et al., 2016). Nonetheless, as Ghisellini et al. (2016) observe, circular economy implementation seems to be in its early stages, focusing on recycling rather than reusing. The "solid waste hierarchy", a system that stresses methods ranging from garbage prevention to waste disposal, strongly impacts waste management.

Indeed, current environmental, social, and economic demands center on the identification of more efficient materials to be used in the transformation industry (Shahbazi et al., 2016), as well as the adoption of a concept based on the value of waste, which should be converted into resources without necessarily reprocessing (Fudala-Ksiazek et al., 2016).

A ZW plan method may solve social challenges when combined with infrastructure, regulatory frameworks, community engagement, and environmentally acceptable treatment technology. The ZW plan supports waste management by encouraging sustainable consumption and production, as well as efficient resource recovery and recycling. The goal of this technique is to reduce waste production to eliminate landfills. Waste reduction, reuse, and recycling are fundamental ideas in ZW's waste management method. Instead of using landfills or incineration, cradle-to-cradle (circular economy) utilizes waste as an input for the next production cycle.

Experts and policymakers evaluate the waste management system's performance using various measures, such as per capita generation rate, collection rate, and recycling rate. These 3R principles are centered on achieving ecological balance, which leads to material and energy savings that benefit the environment. When an item reaches the end of its useful life, it is reused, repaired, sold, or redistributed within the system. Waste products may also be collected, repurposed, and used as inputs, lowering the need for natural resource extraction.

The purpose of this chapter is to offer an overview of ZW, zero landfill (ZL), and the procedures that go with them. The chapter also includes a study and critical review of many techniques for implementing the ZW process and a discussion of the issues that may arise.

2. Sources, Types, and Composition of Solid Waste

Solid waste is described as useless and undesirable items in solid form, i.e., the rubbish generated and abandoned by civilization; as a result, it is one of the most severe environmental issues. Several sectors create massive amounts of solid waste due to human activity. Insights into the type and categorization of wastes aid in determining the need for long-term initiatives such as new technologies, regulations, measures, and policies to reduce landfills' load. The variance in solid waste forms and content, as seen in Fig. 1, is primarily affected by lifestyle, economic status, waste management legislation, and industrial structure. The amount and content of MSW are critical elements in determining the success of trash treatment and management. It is possible to extract energy and resources from solid waste based on its physical composition. Numerous factors, such as socioeconomic status, season, location, and so on, determine the solid waste composition. Understanding the chemical composition of solid waste is critical for evaluating alternative processing and recovery options (Arora et al., 2022). As seen in Fig. 2, solid waste may be classified into two types based on its origin. There are two types of substantial waste: hazardous waste and nonhazardous garbage. Nonhazardous waste includes municipal and agricultural waste. Commercial and household wastes (of the type of sludge waste that cannot be reused and of the type of available waste) are municipal waste. Hazardous waste also includes hospital waste (infectious and noninfectious waste) and industrial waste. Among the many solid waste types are refuse, ashes, and residues, construction and demolition waste, hazardous waste, waste from agriculture, and particular waste types (street cleaning, road litter, and so on).

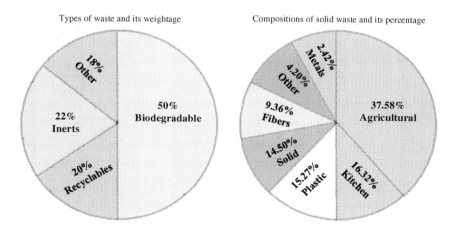

Fig. 1.   Variations in the Content and Form of Solid Waste.

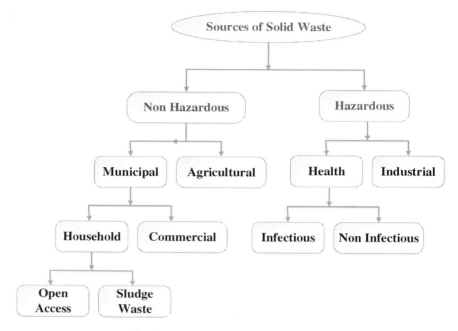

Fig. 2.   Categories of Solid Waste.

3. Waste Management: The Current State and Consequences

Waste management includes waste product collection, storage, disposal, handling, and tracking. The concept of waste management encompasses all materials, whether solids, liquids, gases, or toxic substances, in one category. Additionally, it is intended to minimize the negative environmental consequences associated with best-suited strategies. Because waste characteristics and content vary from source to source, the waste management measure to be implemented depends on the seeds. It is unnecessary to exaggerate the significance of managing solid waste in emerging countries. The management of solid waste is an essential aspect of public health, yet this is underutilized in many impoverished countries. It is due to management system users not properly considering waste management strategies. The waste management system encompasses various duties, from waste product collection to transportation, recycling, and disposal. A waste management system's purpose is to ensure the proper evacuation of wastes from the point of production to the end of processing via suitable disposal. The lack of appropriate waste management poses severe environmental threats to developing countries. In addition to contaminating natural resources, contributing to climate change and global warming, and causing diseases such as typhoid, malaria, cholera, lung infections, and cardiovascular diseases, these diseases adversely affect existing life forms. Apart from vehicular emissions, solid waste management activities emitted around 5% of greenhouse gases in 2016, according to the World Bank report 2020. As industrialization and urbanization increase, rural residents may move into cities for higher-paying jobs (Vij, 2012). The depletion of available resources caused by improper MSWM using conventional waste management strategies poses a critical matter that must be addressed in policy formulation.

Solid waste management encompasses all administrative, financial, legal, planning, and technological tasks associated with a variety of citizen solutions to solid waste challenges. The most frequent means of solid waste disposal around the twentieth century were land dumps, removal, cremation, and till burial. Because of its detrimental environmental impact, open garbage dumping was rapidly recognized as an inefficient waste disposal method and was forbidden in a number of nations. Careless dumping of waste impacts groundwater and soil quality and promotes the spread of diseases transmitted by vectors. Stream dumping, which was formerly popular along the coast, is now illegal because of the damage it does to the surrounding ecosystems. When hazardous substances are dumped into bodies of water, they bio-accumulate and harm all life therein.

Rather than landfilling MSW, more efficient, sustainable, and modern methods need to be used. In terms of cost-effectiveness and convenience, landfills are the best way to dispose of nonrecyclable MSW. Although this has been the case, open landfills have resulted in soil degradation, odor pollution, leachate pollution, and land loss. Because these wastes are not readily digestible and emit greenhouse gases, they seriously threaten the ecosystem. Due to disadvantages, such as public opposition to additional garbage sites, making decisions and adjustments is more critical than ever (Ngamsang & Yuttitham, 2019).

The high calorific value of waste has led to several studies recommending incineration (Singh Parihar & Kumar Gupta, 2020). Incineration became more popular toward the end of the twentieth century because it reduced the waste volume and was more effective at converting energy. The thermal treatment of waste after recycling is one of the most hygienic ways to reduce the amount of waste that has to be disposed of after recycling. During the incineration process, the original amount of garbage can be reduced by up to 85%, and odor and leachate can be eliminated (Karagiannidis et al., 2013). In contrast, MSW incineration requires significant engineering infrastructure. Although trash volumes are reduced after burning, landfills must still dispose of large amounts of waste ash (Parihar & Gupta, 2022).

The majority of energy and material inputs will be converted into valuable products and reusable energy with technological advancements. According to this linear model, natural resources are overused, air, water, and soil are contaminated, and adverse effects are experienced on the environment due to the overuse of natural resources (Yi, 2019). Thus, the question arises: "Where will resources come from in the future?"

4. Framework for Methodology

Significant effort has been made to assess the environmental consequences of various technologies for recovering energy from MSW. The life cycle assessment (LCA) technique is used in most current research, allowing for a complete investigation of possible environmental repercussions (Pires et al., 2011). The most studied waste-to-energy (WtE) methods in the LCA literature are incineration, landfill gas (LFG) collection for energy recovery, and anaerobic digestion (Mayer et al., 2019). In general, waste composition, technological advancement, environmental circumstances, author assumptions, methods, and assumptions (attribution, consequences, system restrictions, and effect examination methodologies) vary significantly throughout the LCA study (Astrup et al., 2015).

4.1 Life Cycle Assessment

In a LCA, any ecological consequences imposed by a system are evaluated (Chen et al., 2019). This approach takes into consideration the entire life cycle of a product, including the origin of raw materials, manufacturing, transportation, usage, and final disposal or recycling.

Local factors such as geographic location, waste kind, source of energy, and demand for recycled commodities often influence policy decisions. The LCA technique may assess the advantages and disadvantages of potential policy changes. LCAs determine the most effective waste management method. An LCA may accurately estimate the environmental impacts and repercussions of products, processes, and services throughout their lifetime (Istrate et al., 2020).

In addition to reducing community pressures and waste management expenses, it reveals opportunities for environmental development. Like material reuse and recycling, incineration with energy recovery decreases dependency on other energy sources and the need for excessive raw material extraction. Biological treatment might also replace inorganic fertilizers and transportation fuel. The LCA of a waste management system involves the following steps:

• Define the goals and scope of the system (define the parameters and boundaries)
• Analyze the inventory (identify inputs and outputs)
• Assess the impact (quantifying environmental impacts)
• Interpret the data (analyzing and comparing all effects and sensitivity analysis)

Existing LCA models for waste management in the region may assist analysts in comprehending how system adjustments affect environmental repercussions via scenario analysis. Suppose significant environmental gains are to be obtained. In that case, LCA must be used to verify that system modifications do not exacerbate environmental deterioration at a later point or time in the Life Cycle.

4.1.1 LCA Benefits

Using a life cycle perspective, all stakeholders (product designers, service providers, political and legislative agencies, and consumers) may make environmentally sound and long-term decisions. LCA's key advantage is that it avoids problem-shifting by accounting for all relevant outcomes. LCA accurately depicts the local system's solid waste, emissions, and energy and material flows (resources). Looking at these maps for various possibilities, such as products or trash management, anybody may find regions for environmental improvement. LCA aims to enable discussion of several environmental concerns rather than just one. LCA methods may give enterprises, governments, and consumers various advantages when making decisions.

5. Municipal Solid Waste Attracts International Attention

MSW has recently received international attention (Govind Kharat et al., 2019). The MSWM is recognized for improving environmental, social, and economic (Turcott Cervantes et al., 2018) sustainability and human health protection. Poor MSWM practices harm the environment, local resources, and social well-being (Yukalang et al., 2017). Each geographical area has its evaluation criteria for trash generation and final disposal location, based on its development and policy goals (Turcott Cervantes et al., 2018). As a result, particular thought must be given to the short- and long-term repercussions of MSWM, both internationally and locally. Prior research has been conducted to identify the MSWM-influencing factors (Singh & Basak, 2018).

Solid waste management in a previous municipality study indicates that MSWM is a complex process that includes waste collecting systems, transfer station sites, treatment procedures, energy recovery, and treatment plant locations. The goal is to balance environmental effectiveness, social acceptability, and economic viability (Morrissey & Browne, 2004).

5.1 Incidences on Policy and Future Directions

The community may confront ongoing challenges in attaining sustainability due to population expansion and the need for upgrading (Ikhlayel, 2018). The quantity of resources recovered from waste streams and used in place of new materials is referred to as resource efficiency. It adds to the alternatives to water, energy, and emissions in MSWM (Deus et al., 2018). It has been established that standardizing contamination reduces MSW chain link contamination and increases recovered site usage (Shekdar, 2009). As a result, the environmental impact is lessened. Environmental deterrence, reuse, recycling, and recovery of landfills decrease the ecological effect. MSWM has gained popularity owing to its economic sufficiency (Arıkan et al., 2017).

The increase in waste generation is a big issue. The average quantity of garbage generated yearly per EU resident is 482 kg (Minelgaitė & Liobikienė, 2019). According to the authors, waste management operations' success in minimizing rubbish buildup or, better still, waste creation is mainly determined by public acceptability and behavior (Pandey et al., 2018; Wan et al., 2018).

The EU nations with the most significant waste generation rates are also economically developed. Developing strategies for waste reduction needs to consider the broader objective of severing the link between trash generation and economic development.

6. A Zero-Waste Strategy for Reducing Energy Consumption

MSWM has gained popularity due to its impact on economic growth, environmental preservation, and public health. Green supply chain management (GSCM) aims to handle end-of-life items (or waste products) after their useful lives. An essential first step in establishing and maintaining a clean environment is to address the global waste problem (Iqbal et al., 2020). As a result, GSCM urges businesses to decrease waste to preserve the environment. The goal of GSCM is to decrease product waste while also conserving energy. Supply chain managers now need to reduce their energy use due to rising energy costs and the effects of energy consumption on the environment. Future generations need energy supplies that balance production and consumption.

Given the scarcity of available energy, it is critical to minimize global energy use and create renewable energy sources. Energy is utilized throughout the supply chain to develop and distribute consumer goods. Following specific dispatching rules and production sequences may reduce energy usage in such a supply chain (Mouzon et al., 2007). Clients get deliveries of packaged consumer items such as food. Most consumer goods rubbish is classified as "product waste and packaging junk." Organic food waste may be digested and converted into secondary products like fertilizer, animal feed, or biofuel.

6.1 Zero Waste Concept

The concept of ZW encompasses a broad concept that encourages waste management like sustainable consumption and production, effective resource recovery, and recycling. This strategy tries to limit garbage production to eliminate landfills.

The concept has only been adopted in a few countries, usually industrialized ones with enough financing. The growth in waste output is due to increased resource use. As per capita rubbish generation rises, the Zero Waste goal remains elusive. ZW's strategy adheres to the core principles of a sustainable waste management system: reduce, reuse, and recycle. Rather than using landfills and incineration to dispose of waste, we will be using cradle-to-cradle (circular economics), which utilizes waste as an input for a subsequent production cycle. The development of ecological balance is at the core of these 3R principles, resulting in material and energy savings that will benefit the environment. When an object reaches the end of its useful life, it is reused, repaired, sold, or redistributed within the system. Material from the waste stream may also be collected, repurposed, and used as inputs, minimizing the demand for natural resource extraction.

ZW's core strategy is waste avoidance and deterrent rather than garbage treatment and disposal. Waste prevention requires broad social awareness, understanding of waste, and innovative industrial and economic strategies (Curran & Williams, 2012). Sustainable and ethical shopping habits hamper waste avoidance. The ZW strategy, when combined with infrastructure, regulatory frameworks, community engagement, and environmentally acceptable treatment technologies, has the potential to tackle social concerns. Experts and policymakers use a range of measures to evaluate the efficacy of the waste management system, including per capita creation rate, collection rate, and recycling rate. Fig. 3 shows how the traditional technology has given way to zero waste technology. It demonstrates how energy-efficient supply networks and product life cycles preserve the maximum amount of value and quality for the longest possible period (Arora et al., 2022).

6.2 Zero Waste and Zero Landfills – The Key to Success

Increasing urbanization and population growth have dramatically raised the worldwide waste generation. Because all actions generate trash, examining how to do so may encourage us to reevaluate how we identify, create, and dispose of it. ZW designs prioritize waste reduction and material reuse above material recycling. Waste is often used to describe abandoned and discarded items in landfills (Hamid et al., 2020).

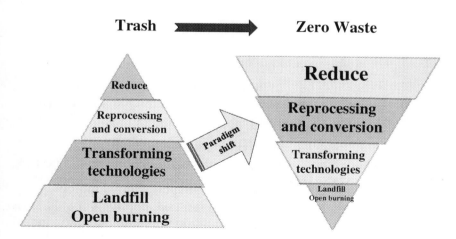

Fig. 3.   Demonstration of the Paradigm Transition From Old Technologies to Zero Waste Technologies.

The human population has more than doubled in the previous four decades, putting strain on demand and supply networks. Furthermore, rising per capita waste generation and inadequate waste management systems substantially impact the availability of essential resources (Gaur et al., 2022). Natural resources are extracted and converted into afterward discarded or burnt goods. This linear paradigm is incompatible with a sustainable process. It has significant adverse environmental consequences due to the overuse of limited natural resources and the degradation of the quality of local ecological resources. Following that, resource recovery and reverse logistics (circular economy) emerged. Cradle-to-cradle waste management (lower use of primary resources) is more sustainable than cradle-to-grave waste management since trash from one product may be utilized as an input for another. As a result, landfills and incineration are no longer the only MSW possibilities. Understanding the reverse logistics notion may be aided by seeing how nature produces no waste (ZW). Natural ecosystems are distinguished by symbiotic and cooperative interactions in which the results of one activity serve as the starting point for another. Circular material movement strengthens and maintains the system.

6.3 Implementing the Zero-Waste Philosophy: Benefits and Challenges

The ideal ZW system maintains continual resource circulation (circular design). As a result, ZW programs reduce the demand for virgin materials in manufacturing new goods and reduce landfill and incinerator use, minimizing the risks to humans and the planet. The ZW strategy reduces energy consumption for material creation and collection by eliminating the need for virgin material extraction and processing. The benefits of ZW are classified into four categories.

- Community benefits
  Incentives for public participation in implementing ZW, changes in consumer behavior, enhanced employment possibilities, and changes in waste generating and disposal habits reduce the risk to public health.
- Financial and economic benefits
  As expenses are decreased, profitability rises. It is possible to avoid environmental restoration costs and losses resulting from inefficient methods by increasing revenue from recycled materials and enhancing cash flow due to system-related employment.
- Environmental benefits
  The use of virgin raw resources is declining. Reduced usage of potentially hazardous substances in products minimizes CO_2 emissions, waste production, and adverse effects. It is feasible to generate power from rubbish and sell carbon credits. Utilization in sanitary landfills has grown while load has decreased – reduced energy usage resulted from more eco-friendly production and recycling practices and more incredible dedication to environmental preservation.

- Industry benefits
  Industrial symbiosis refers to a company that provides waste to another company and vice versa. In this case, a stable supply chain is created.

6.3.1 Zero-Waste Strategies to Improve Waste Management and Recycling

Municipal governments provide a public service of collecting and disposing of MSW, which is impacted by the cost, environmental concerns, and complexity of MSW management (Pérez-López et al., 2016). MSW may be handled in two ecologically responsible ways: trash reduction and recycling via garbage sorting (Astrup et al., 2015). The demographic and socioeconomic features of the population serviced (Czajkowski et al., 2015), geographical and structural variables (Mazzanti et al., 2011), operational characteristics (Guerrini et al., 2017), and government characteristics (Gaeta et al., 2017) all influence how successfully waste management firms work. Typically, these parameters are quantified regarding municipal garbage generation, recycling rate, and cost efficiency.

Municipal rubbish output increases when privately held enterprises manage municipal garbage services and when taxable per capita income is lower (Romano et al., 2019). Recycling has been the subject of several studies (Gu et al., 2017). Dodbiba et al. (2008) examined waste polymers from obsolete TVs using mechanical recycling and energy recovery. Materials and energy recovery were investigated by Rigamonti et al. (2014).

It is feasible to efficiently minimize MSW by incinerating waste and producing power simultaneously (Yazdani et al., 2020). The most challenging aspect of this operation is dealing with waste products such as bottom ash, fly ash, and lime required for air pollution control (APC) (Ashraf et al., 2019).

Municipalities all across the globe generate billions of tons of solid trash each year. Most MSW is disposed either by landfilling or burning. In comparison to landfilling, incineration reduces waste volume and bulk by 90% and 70%, respectively. It also allows for the recovery of energy in the form of electricity (Joseph et al., 2018).

6.3.2 A Zero-Waste Management Strategy for the Industrial Sector

Various methodologies, methods, tools, and ideas have been employed in the area of zero waste management (ZWM) strategies for products to solve waste management and resource efficiency concerns from raw material extraction through final disposal. The four steps that comprise the fundamental approaches indicated for adopting ZW in solid waste management are as follows: the first level, concerned with eco-design, cutting-edge technologies, LCA, and product sustainability, may be characterized by using energy and environmental analytical methods. These efforts will cut resource and energy usage while improving product functionality and making producers' commitments extremely obvious. The second level, which is concerned with manufacturing processes, focuses mainly on the development of goods and manufacturing methods that are consistent with ecological cycles and clean manufacturing strategies. It aims to boost productivity while reducing pollution and waste. Examining the flow of resources and energy through an organization may try to uncover solutions to decrease waste and pollution from industrial activities. The third stage includes environmental awareness campaigns and eco-labeling (selling and usage). Without appropriate ecological awareness, achieving ZW targets would be difficult. Because the existing spending pattern is unsustainable and cannot be sustained continuously, it is critical to recognize this and take necessary action. In reality, behavioral improvements must account for 25% of the decrease in emissions. An assessment agency certifies a product, process, or management system to fulfill particular environmental standards as part of ecolabel campaigns that encourage customers to purchase ecologically friendly items. The most important aspect of the fourth level, also known as the end-of-life phase, is the establishment of a successful managed environmental program that organizes, executes, and controls activities, such as pollution prevention and material recovery, that aim to improve environmental performance. It is a problem-solving strategy that assists organizations in meeting their environmental obligations and performance objectives while systematically managing their operations, goods, and services.

6.3.3 Construction of Zero-Waste Buildings

The construction industry significantly contributes to a country's GDP. The construction sector impacts people's social, economic, and environmental well-being because of its links to nation-building, substantial use of labor and equipment, and resource consumption. Every nation must have social and economic advantages since they help it thrive. However, it also offers environmental concerns, so each building must balance social, economic, and ecological issues equally.

Alternative construction techniques must be considered, including the proper selection of raw materials, building structure, the building's duration, use, adaptability during that lifespan, and end-of-life utilization.

Several studies have been conducted on the material phase of a structure and its implications for raw material extraction and manufacture (Ashok Kumar et al., 2022).

The sustainable building aims for maximum resource efficiency, flexibility to meet future user needs, quality of life and customer satisfaction, and desirable ecological and social surroundings.

Sustainable building strives to foster economic development while emphasizing social and environmental integrity via several strategies utilized throughout a project's life cycle.

6.3.3.1 Demolition and Construction Waste. Most structures are built with a few decades of existence in mind, and potential demolition is seldom addressed. The construction industry's generation of so much rubbish has a severe impact on society, the economy, and the environment. The world's resources are finite. Garbage disposal in landfills will become more difficult as the world's population grows. A variety of readily accessible materials are used in the construction business. What happens to destroyed structures is not considered in the construction industry. The majority of demolition waste is disposed of in landfills. Despite heightened recycling rates in certain nations, most recovered materials are downcycled. This waste of resources and commodities is unsustainable if the global economy and population continue to grow. Waste is any substance that cannot be utilized for its intended purpose. The major consequences of this were material contamination or mixing. The original content must be restored, which requires a significant amount of time and money. To guarantee a more sustainable use of resources, the business sector should strive toward waste-free operations. This technique's purpose is to construct a closed loop for all materials. Recycling, repairing materials for use in technology, or allowing them to degrade naturally are all required. Using the design for deconstruction method, you may achieve ZW. Deconstruction is the methodical disassembly of a structure after the end of its useful life. The dismantled pieces might be used or utilized in fresh construction (Akhtar & Sarmah, 2018). To prevent downcycling, all materials should be returned to the biosystem or technological system. The material flow diagram during demolition and construction is shown in Fig. 4.

7. Discussion

International programs aimed at achieving more sustainable MSWM systems include waste-to-energy (Abd Kadir et al., 2013), waste-to-zero waste (Song et al., 2015), smart waste systems (Fujii et al., 2014), and ZW (Fuss et al., 2018). Policymakers favor the ZW concept because it supports resource recovery and recycling optimization, sustainable consumption and production, and avoiding trash incineration and disposal. Waste management firms have used and comprehended the ZW concept in various ways (Li & Du, 2015). Landfills and incineration are prohibited under ZW principles. At the same time, some studies

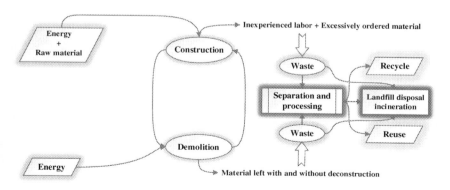

Fig. 4.   The Schematic of Material Flow Diagram During Demolition and Construction.

argue that waste-to-energy technologies such as incineration may be employed as a waste recovery approach to achieve ZW goals (Premalatha et al., 2013). They also say that the ZW concept should be extended to all situations (Kabirifar et al., 2020). Using the ZWM framework, decision-makers may be able to assess technology options for enhancing waste minimization, sorting, and recycling throughout the waste value chain. Three tactics are required to create a smart city based on ZW: preventing waste, suitable trash disposal, and value recovery (Esmaeilian et al., 2018). In the industrial and waste management sectors, the current adoption of ZWM systems may provide a variety of challenges. There are chances to reduce trash production and recover value from the vast quantities of incinerated waste. Still, a number of obstacles must be overcome before the ZWM themes' technologies can be applied. Depending on the technology, planning for disassembly might affect previously consumed items within the concept of ZW design. Modular building construction has already been shown to be incredibly beneficial in building construction. There is already a commitment from the HDB (Housing Development Board) to produce 35% of new housing complexes using factory-produced modular units rather than on-site units (Kerdlap et al., 2019).

To efficiently manage materials, it is also vital to evaluate the reusability and recyclable qualities of a building's overall breakdown structures of created material to its independent components and, eventually, to primary raw material. Sustainable development aims to integrate the environment, society, and economy. Building waste management must balance economic, social, and environmental factors. The use of additive manufacturing technology is a continual research and development subject, which will reduce the production of waste metals, polymers, and construction materials while increasing the energy efficiency of manufacturing processes.

Intelligent waste audit and reduction solutions for waste data management and benchmarking are conceivable on the software side. The physical method of collecting trash data is a technical challenge that must be overcome.

8. Conclusions

MSWM has been acknowledged for its importance in municipal growth and environmental conservation. More effective waste management is required now more than ever with rising garbage output, high landfilling rates, and poor waste reduction rates. Reverse logistics (circular economy) is an exciting concept in a society with limited resources. Excessive use of virgin resources is unsustainable, environmentally harmful, and costly. Natural systems in which trash from one company or people becomes nourishment for another inspired the notion of Reuse, Recycle, and Reduce (or "zero waste"). ZW is not only an innovative approach to solving the world's waste issue but is also tough to attain. For ZW to become a reality, it requires long-term commitment and active engagement among all important players, including manufacturers, end users, and government officials. In addition to environmental advantages, adopting ZW techniques seems to be advantageous for establishing new business prospects and economic momentum.

The ZW technique requires a cyclical material flow in which resources are valued. Closed loop systems develop, use, and reuse resources rather than burning and burying them. In reverse logistics, a well-managed resource recycling strategy can protect and utilize our resources while saving money and reducing the impact on landfills and natural resources. Even though just a tiny percentage of the industrialized world has accepted the ZW concept, it is expected to gain popularity in developing nations shortly, benefiting both their economy and the environment. A waste-free planet is a significant objective that can only be realized with the participation of all stakeholders and a change in their behavior.

9. Further Research Suggestions

There are so many concerns in this field that the present investigation can't cover them all. In light of the outcomes of this investigation, it has been concluded that the following items warrant additional examination.

- The concept could be expanded in further research to a broader setting. More study into ZW solutions across various sectors is advised in light of the conceptual framework presented in this chapter.
- Future research might focus on the new evaluation methods such as Emergy and the hybrid strategies. The modern studies indicated that integrating LCA, Emergy, and exergy could help make an appropriate decision.
- The second issue is deciding on dependent variables for implementing resource efficiency measures. Critical information about the intensity of resource efficiency measures is lost when process reengineering or waste management activities are viewed as binary variables. Future research could investigate the significance of the various relationships revealed in this study using long-term data on resource efficiency measures.

Acknowledgments

This research was supported by Brain Pool program funded by the Ministry of Science and ICT through the National Research Foundation of Korea (NRF-2022H1D3A2A02090885).

References

Abd Kadir, S. A. S., Yin, C.-Y., Rosli Sulaiman, M., Chen, X., & El-Harbawi, M. (2013). Incineration of municipal solid waste in Malaysia: Salient issues, policies and waste-to-energy initiatives. *Renewable and Sustainable Energy Reviews, 24*, 181–186.

Akhtar, A., & Sarmah, A. K. (2018). Construction and demolition waste generation and properties of recycled aggregate concrete: A global perspective. *Journal of Cleaner Production, 186*, 262–281.

Arıkan, E., Şimşit-Kalender, Z. T., & Vayvay, Ö. (2017). Solid waste disposal methodology selection using multi-criteria decision making methods and an application in Turkey. *Journal of Cleaner Production, 142*, 403–412.

Aristei, D., & Gallo, M. (2021). The role of external support on the implementation of resource efficiency actions: Evidence from European manufacturing firms. *Sustainability, 13*.

Arora, S., Sethi, J., Rajvanshi, J., Sutaria, D., & Saxena, S. (2022). Developing "zero waste model" for solid waste management to shift the paradigm toward sustainability. In *Handbook of solid waste management* (pp. 345–364). Springer.

Ashok Kumar, G., Sanjay Kumar, S., & Hazi, A. (2022). *Advances in construction materials and sustainable environment.* Springer.

Ashraf, M. S., Ghouleh, Z., & Shao, Y. (2019). Production of eco-cement exclusively from municipal solid waste incineration residues. *Resources, Conservation and Recycling, 149*, 332–342.

Astrup, T. F., Tonini, D., Turconi, R., & Boldrin, A. (2015). Life cycle assessment of thermal Waste-to-Energy technologies: Review and recommendations. *Waste Management, 37*, 104–115.

Bodas-Freitas, I.-M., & Corrocher, N. (2019). The use of external support and the benefits of the adoption of resource efficiency practices: An empirical analysis of European SMEs. *Energy Policy, 132*, 75–82.

Brunner, P. H., & Fellner, J. (2016). Setting priorities for waste management strategies in developing countries. *Waste Management & Research: The Journal for a Sustainable Circular Economy, 25*, 234–240.

Campitelli, A., Cristóbal, J., Fischer, J., Becker, B., & Schebek, L. (2019). Resource efficiency analysis of lubricating strategies for machining processes using life cycle assessment methodology. *Journal of Cleaner Production, 222*, 464–475.

Chen, Y., Cui, Z., Cui, X., Liu, W., Wang, X., Li, X., & Li, S. (2019). Life cycle assessment of end-of-life treatments of waste plastics in China, Resources. *Conservation & Recycling, 146*, 348–357.

Coban, A., Ertis, I. F., & Cavdaroglu, N. A. (2018). Municipal solid waste management via multi-criteria decision making methods: A case study in Istanbul, Turkey. *Journal of Cleaner Production, 180*, 159–167.

Curran, T., & Williams, I. D. (2012). A zero waste vision for industrial networks in Europe. *Journal of Hazardous Materials, 207–208*, 3–7.

Czajkowski, M., Hanley, N., & Nyborg, K. (2015). Social norms, morals and self-interest as determinants of pro-environment behaviours: The case of household recycling. *Environmental and Resource Economics, 66*, 647–670.

Deus, R. M., Bezerra, B. S., & Battistelle, R. A. G. (2018). Solid waste indicators and their implications for management practice. *International journal of Environmental Science and Technology, 16*, 1129–1144.

Dodbiba, G., Takahashi, K., Sadaki, J., & Fujita, T. (2008). The recycling of plastic wastes from discarded TV sets: Comparing energy recovery with mechanical recycling in the context of life cycle assessment. *Journal of Cleaner Production, 16*, 458–470.

Esmaeilian, B., Wang, B., Lewis, K., Duarte, F., Ratti, C., & Behdad, S. (2018). The future of waste management in smart and sustainable cities: A review and concept paper. *Waste Management, 81*, 177–195.

Flachenecker, F., & Rentschler, J. (2018). *Investing in resource efficiency.* Springer.

Fudala-Ksiazek, S., Pierpaoli, M., Kulbat, E., & Luczkiewicz, A. (2016). A modern solid waste management strategy – The generation of new by-products. *Waste Management, 49*, 516–529.

Fujii, M., Fujita, T., Ohnishi, S., Yamaguchi, N., Yong, G., & Park, H.-S. (2014). Regional and temporal simulation of a smart recycling system for municipal organic solid wastes. *Journal of Cleaner Production, 78*, 208–215.

Fuss, M., Vasconcelos Barros, R. T., & Poganietz, W.-R. (2018). Designing a framework for municipal solid waste management towards sustainability in emerging economy countries – An application to a case study in Belo Horizonte (Brazil). *Journal of Cleaner Production, 178*, 655–664.

Gaeta, G. L., Ghinoi, S., & Silvestri, F. (2017). Municipal performance in waste recycling: An empirical analysis based on data from the Lombardy region (Italy). *Letters in Spatial and Resource Sciences, 10*, 337–352.

Gaur, A., Gurjar, S. K., & Chaudhary, S. (2022). Circular system of resource recovery and reverse logistics approach: Key to zero waste and zero landfill. In *Advanced organic waste management* (pp. 365–381). Elsevier.

Ghisellini, P., Cialani, C., & Ulgiati, S. (2016). A review on circular economy: The expected transition to a balanced interplay of environmental and economic systems. *Journal of Cleaner Production, 114*, 11–32.

Govind Kharat, M., Murthy, S., Jaisingh Kamble, S., Raut, R. D., Kamble, S. S., & Govind Kharat, M. (2019). Fuzzy multi-criteria decision analysis for environmentally conscious solid waste treatment and disposal technology selection. *Technology in Society, 57*, 20–29.

Guerrero, L. A., Maas, G., & Hogland, W. (2013). Solid waste management challenges for cities in developing countries. *Waste Management, 33*, 220–232.

Guerrini, A., Carvalho, P., Romano, G., Cunha Marques, R., & Leardini, C. (2017). Assessing efficiency drivers in municipal solid waste collection services through a non-parametric method. *Journal of Cleaner Production, 147*, 431–441.

Gu, F., Guo, J., Zhang, W., Summers, P. A., & Hall, P. (2017). From waste plastics to industrial raw materials: A life cycle assessment of mechanical plastic recycling practice based on a real-world case study. *Science of the Total Environment, 601–602*, 1192–1207.

Hamid, S., Skinder, B. M., & Bhat, M. A. (2020). Zero waste. In *Innovative waste management technologies for sustainable development* (pp. 134–155). IGI Global.

Ikhlayel, M. (2018). Indicators for establishing and assessing waste management systems in developing countries: A holistic approach to sustainability and business opportunities. *Business Strategy & Development, 1*, 31–42.

İncekara, M. (2022). Determinants of process reengineering and waste management as resource efficiency practices and their impact on production cost performance of Small and Medium Enterprises in the manufacturing sector. *Journal of Cleaner Production, 356*, 131712.

Iqbal, M. W., Kang, Y., & Jeon, H. W. (2020). Zero waste strategy for green supply chain management with minimization of energy consumption. *Journal of Cleaner Production, 245*, 118827.

Istrate, I.-R., Iribarren, D., Gálvez-Martos, J.-L., & Dufour, J. (2020). Review of life-cycle environmental consequences of waste-to-energy solutions on the municipal solid waste management system. *Resources, Conservation and Recycling, 157*, 104778.

Joseph, A., Snellings, R., Van den Heede, P., Matthys, S., & De Belie, N. (2018). The use of municipal solid waste incineration ash in various building materials: A Belgian point of view. *Materials, 11*, 141.

Kabirifar, K., Mojtahedi, M., Wang, C., & Tam, V. W. Y. (2020). Construction and demolition waste management contributing factors coupled with reduce, reuse, and recycle strategies for effective waste management: A review. *Journal of Cleaner Production, 263*, 121265.

Karagiannidis, A., Kontogianni, S., & Logothetis, D. (2013). Classification and categorization of treatment methods for ash generated by municipal solid waste incineration: A case for the 2 greater metropolitan regions of Greece. *Waste Management, 33*, 363–372.

Kerdlap, P., Low, J. S. C., & Ramakrishna, S. (2019). Zero waste manufacturing: A framework and review of technology, research, and implementation barriers for enabling a circular economy transition in Singapore. *Resources, Conservation and Recycling, 151*, 104438.

Khatib, I. A. (2011). Municipal solid waste management in developing countries: Future challenges and possible opportunities. *Integrated Waste Management, 2*, 35–48.

Li, R. Y. M., & Du, H. (2015). Sustainable construction waste management in Australia: A motivation perspective. In *Construction safety and waste management* (pp. 1–30). Springer.

Mayer, F., Bhandari, R., & Gäth, S. (2019). Critical review on life cycle assessment of conventional and innovative waste-to-energy technologies. *Science of the Total Environment, 672*, 708–721.

Mazzanti, M., Montini, A., & Nicolli, F. (2011). Embedding landfill diversion in economic, geographical and policy settings. *Applied Economics, 43*, 3299–3311.

Minelgaitė, A., & Liobikienė, G. (2019). Waste problem in European Union and its influence on waste management behaviours. *Science of the Total Environment, 667*, 86–93.

Morrissey, A. J., & Browne, J. (2004). Waste management models and their application to sustainable waste management. *Waste Management, 24*, 297–308.

Mouzon, G., Yildirim, M. B., & Twomey, J. (2007). Operational methods for minimization of energy consumption of manufacturing equipment. *International Journal of Production Research, 45*, 4247–4271.

Ngamsang, T., & Yuttitham, M. (2019). Vulnerability assessment of areas allocated for municipal solid waste disposal systems: A case study of sanitary landfill and incineration. *Environmental Science and Pollution Research, 26*, 27239–27258.

Özbuğday, F. C., Fındık, D., Metin Özcan, K., & Başçı, S. (2020). Resource efficiency investments and firm performance: Evidence from European SMEs. *Journal of Cleaner Production, 252*.

Pandey, R. U., Surjan, A., & Kapshe, M. (2018). Exploring linkages between sustainable consumption and prevailing green practices in reuse and recycling of household waste: Case of Bhopal city in India. *Journal of Cleaner Production, 173*, 49–59.

Parihar, N. S., & Gupta, A. K. (2022). Effect of curing on compressive and shear strength parameters of liming waste ash stabilized expansive soil. In *Advances in construction materials and sustainable environment* (pp. 1035–1046). Springer.

Pérez-López, G., Prior, D., Zafra-Gómez, J. L., & Plata-Díaz, A. M. (2016). Cost efficiency in municipal solid waste service delivery. Alternative management forms in relation to local population size. *European Journal of Operational Research, 255,* 583–592.

Pietzsch, N., Ribeiro, J. L. D., & de Medeiros, J. F. (2017). Benefits, challenges and critical factors of success for zero waste: A systematic literature review. *Waste Management, 67,* 324–353.

Pires, A., Martinho, G., & Chang, N.-B. (2011). Solid waste management in European countries: A review of systems analysis techniques. *Journal of Environmental Management, 92,* 1033–1050.

Premalatha, M., Tauseef, S. M., Abbasi, T., & Abbasi, S. A. (2013). The promise and the performance of the world's first two zero carbon eco-cities. *Renewable and Sustainable Energy Reviews, 25,* 660–669.

Rigamonti, L., Grosso, M., Møller, J., Martinez Sanchez, V., Magnani, S., & Christensen, T. H. (2014). Environmental evaluation of plastic waste management scenarios. *Resources, Conservation and Recycling, 85,* 42–53.

Romano, G., Rapposelli, A., & Marrucci, L. (2019). Improving waste production and recycling through zero-waste strategy and privatization: An empirical investigation. *Resources, Conservation and Recycling, 146,* 256–263.

Shahbazi, S., Wiktorsson, M., Kurdve, M., Jönsson, C., & Bjelkemyr, M. (2016). Material efficiency in manufacturing: Swedish evidence on potential, barriers and strategies. *Journal of Cleaner Production, 127,* 438–450.

Shekdar, A. V. (2009). Sustainable solid waste management: An integrated approach for Asian countries. *Waste Management, 29,* 1438–1448.

Singh Parihar, N., & Kumar Gupta, A. (2020). Chemical stabilization of expansive soil using liming leather waste ash. *International Journal of Geotechnical Engineering, 15,* 1008–1020.

Singh, A., & Basak, P. (2018). Economic and environmental evaluation of municipal solid waste management system using industrial ecology approach: Evidence from India. *Journal of Cleaner Production, 195,* 10–20.

Sisto, R., Sica, E., Lombardi, M., & Prosperi, M. (2017). Organic fraction of municipal solid waste valorisation in southern Italy: The stakeholders' contribution to a long-term strategy definition. *Journal of Cleaner Production, 168,* 302–310.

Smol, M., Kulczycka, J., & Kowalski, Z. (2016). Sewage sludge ash (SSA) from large and small incineration plants as a potential source of phosphorus – Polish case study. *Journal of Environmental Management, 184,* 617–628.

Soltani, A., Hewage, K., Reza, B., & Sadiq, R. (2015). Multiple stakeholders in multi-criteria decision-making in the context of municipal solid waste management: A review. *Waste Management, 35,* 318–328.

Song, Q., Li, J., & Zeng, X. (2015). Minimizing the increasing solid waste through zero waste strategy. *Journal of Cleaner Production, 104,* 199–210.

Su, J.-P., Chiueh, P.-T., Hung, M.-L., & Ma, H.-W. (2007). Analyzing policy impact potential for municipal solid waste management decision-making: A case study of Taiwan. *Resources, Conservation and Recycling, 51,* 418–434.

Turcott Cervantes, D. E., López Martínez, A., Cuartas Hernández, M., & Lobo García de Cortázar, A. (2018). Using indicators as a tool to evaluate municipal solid waste management: A critical review. *Waste Management, 80,* 51–63.

Vij, D. (2012). Urbanization and solid waste management in India: Present practices and future challenges. *Procedia-Social and Behavioral Sciences, 37*, 437–447.

Wan, C., Shen, G. Q., & Choi, S. (2018). Differential public support for waste management policy: The case of Hong Kong. *Journal of Cleaner Production, 175*, 477–488.

Yang, J., Firsbach, F., & Sohn, I. (2022). Pyrometallurgical processing of ferrous slag "co-product" zero waste full utilization: A critical review. *Resources, Conservation and Recycling, 178*.

Yazdani, S., Salimipour, E., & Moghaddam, M. S. (2020). A comparison between a natural gas power plant and a municipal solid waste incineration power plant based on an emergy analysis. *Journal of Cleaner Production, 274*.

Yi, S. (2019). Resource recovery potentials by landfill mining and reclamation in South Korea. *Journal of Environmental Management, 242*, 178–185.

Yukalang, N., Clarke, B., & Ross, K. (2017). Barriers to effective municipal solid waste management in a rapidly urbanizing area in Thailand. *International Journal of Environmental Research and Public Health, 14*.

Yu, H., & Solvang, W. D. (2017). A multi-objective location-allocation optimization for sustainable management of municipal solid waste. *Environment Systems and Decisions, 37*, 289–308.

Zaman, A. U. (2014). Identification of key assessment indicators of the zero waste management systems. *Ecological Indicators, 36*, 682–693.

Chapter 8

Adoption of Green Building Practices in Kenya: A Case of Kakamega Municipality

Edwin K. Kanda, Elizabeth Lusweti, Francis N. Ngugi, John M. Irungu, Bernard O. Omondi and Samuel G. Waweru

Abstract

Green building (GB) is an important aspect of sustainable development that advocates for practices that enhance the health and well-being of the occupants and communities with minimal impact on the environment. The adoption of the GB concept is low in most developing countries. This study aimed at assessing the extent of the adoption of GB concepts in Kakamega municipality as a case study. The study sampled 64 respondents which consisted of 15 professionals in the building industry and 49 commercial and residential buildings within the municipality. The results indicated that at least 80% of the professionals were aware of GB concepts of water efficiency, energy efficiency, sustainable materials, sustainable site practices, and indoor environmental quality (IEQ). Water efficiency practices that were widely adopted in commercial and residential buildings were rainwater harvesting and the use of efficient plumbing fixtures. The energy-saving measures adopted were switching off appliances when not in use, use of energy-saving bulbs, and solar energy. The majority of the respondents felt that the indoor air quality was fair. Resource reuse was not widely adopted which could be attributed to a lack of policy or legislation. Low impact development (LID) design of vegetated storm conveyance and rain gardens (bioretention and porous/permeable pavements) was adopted in 42.9% of the buildings. Research into the use of sustainable materials is recommended for wider application as a GB concept. The study recommends the promotion of GB through incentives to enable wider adoption among the owners and developers. Furthermore, GB legislation and certification programs need to be adopted in Kenya. This study was largely based on Leadership in Energy and Environmental Design (LEED) criteria and thus further studies are required on other assessment tools and methods.

Pragmatic Engineering and Lifestyle, 153–169
doi:10.1108/978-1-80262-997-220231008

Keywords: Green building; Kenya; Kakamega; sustainable development; indoor air quality; solar energy

1. Introduction

Sustainable development in the construction industry requires the adoption of practices that conserve resources through efficient and clean technologies. Green building (GB) is one aspect of sustainable development since the concept advocates for the adoption of practices that are considered environmentally friendly. There are various concepts/practices which are synonymous with GB such as sustainable construction and sustainable design. This study adopts the definition by Robichaud and Anantatmula (2011) where GB is both a philosophy as well as associated project and construction management practices that seek to: (1) minimize or eliminate impacts on the environment, natural resources, and non-renewable energy sources to promote the sustainability of the built environment; (2) enhance the health, well-being, and productivity of occupants and communities; (3) cultivate economic development and financial returns for developers and communities, and (4) apply life cycle approaches to community planning and development. GBs are constructed based on the principles of sustainable construction, which addresses the ecological, social, and economic issues of a building in the context of its community (Hwang & Tan, 2012). Therefore, GB adopts the

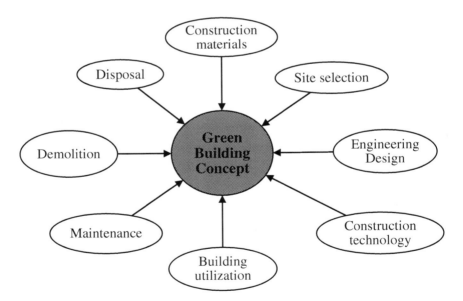

Fig. 1.   Focus Areas of a Green Building. *Source:* Authors' illustration.

sustainability principles of construction (see Fig. 1) in terms of extraction of construction materials, engineering design and development, site selection, construction methods and technologies, operation (utilization), maintenance, and demolition including disposal. These elements require research and innovation, especially in the choice of construction materials, engineering design, construction methods, and disposal mechanisms.

There is a growing awareness that sustainable development has the potential to positively impact the environment and thus pushing the GB concept to the forefront (Robichaud & Anantatmula, 2011). GBs have the potential of reducing greenhouse gas emissions (Olanipekun et al., 2017) and thus can be considered to be a climate-smart concept. According to Choi (2009), in the United States, buildings accounted for almost 40% of the total energy consumption and around 38% of the total carbon dioxide gas emissions.

Adoption of the philosophy starts with building design where the engineers and architects conceptualize the elements that constitute the GB and they must be able to persuade or convince the project owners (clients) that the initial high costs will be economically attractive in the long term. The next category is the contractors who will put up the building using sustainable construction techniques and materials. Finally, the occupants and owners must feel comfortable in the GB. According to Mesthrige and Kwong (2018), end users' needs and comfort and their perspectives play an important role in determining and rating a GB.

Therefore, there must be incentives for the actors above to adopt GB practices. According to Olubunmi et al. (2016), these incentives can be external or internal with the former being offered by an external agency, for example, government in the form of financial (taxes) and non-financial (floor-to-area density, technical assistance, expedited permitting, regulatory relief, marketing assistance, etc.) and the latter being intrinsic motivation by project owners or stakeholders. External drivers of incentives are sometimes referred to as the top-down approach and include legislation and policies and other incentives from national and local governments (Ahn et al., 2013). Project owners are motivated to go "green" for several reasons such as personal values, a sense of corporate responsibility, marketing angles, a desire to keep up with the standard set by their business type, and economics (Bornais, 2012). Also, market forces can drive project owners to adopt sustainable construction (Ahn et al., 2013). Behavioral and cultural factors are also very important in GB concept adoption, and thus awareness levels of stakeholders (e.g., clients, designers, contractors, and end users) on concepts of sustainable buildings are needed (Zuo & Zhao, 2014).

There are many tools for assessing and certifying GBs. The most notable ones include Leadership in Energy and Environmental Design (LEED, United States), British Research Establishment Environmental Assessment Method (BREEAM, United Kingdom), Green Building Council of Australia Green Star (GBCA, Australia), Green Mark Scheme (Singapore), German Sustainable Building Council DGNB (Germany), Comprehensive Assessment System for Built Environment Efficiency (CASBEE, Japan), Pearl Rating System for Estidama (Abu Dhabi Urban Planning Council), Hong Kong Building Environmental Assessment Method (HK BEAM), and Green Building Index (Malaysia) which have been developed and used for assessing GB developments (Zuo & Zhao, 2014). These tools assess buildings using several criteria which are categorized into green aspects. For example, LEED uses six areas to certify a building as green. These areas are sustainable site development, water efficiency, energy efficiency, use and reuse of resources and materials, indoor environmental quality (IEQ), and innovation and design (Pitts & Jackson, 2008). The LEED features for the GB assessment apply to developing countries like Kenya (Khaemba & Mutsune, 2014). BREEAM focuses on energy use, transportation, water, ecology, land use, materials, pollution, and health and well-being; Green Mark Scheme covers energy efficiency, water efficiency, environmental protection, IEQ, and innovation while Green Building Index considers energy efficiency, IEQ, sustainable site and management, materials and resources, and water efficiency (Komolafe et al., 2016).

The rate of implementation of GB concepts is low in low- and medium-income countries despite its benefits to the environment. In Vietnam, Nguyen et al. (2017) noted that slow policy-making and lack of comprehensive policy, lack of awareness and misconceptions about the GB are some of the barriers hindering adoption. Lack of awareness and lack of enabling environment in the form of policy or legislation are some of the main factors hindering the adoption of GB construction in Nigeria (Dahiru et al., 2014). The high cost of GB technologies coupled with inadequate financing schemes and lack of government incentives were the main barriers to the adoption of GB initiatives in Ghana (Chan et al., 2018).

In Kenya, aspects of GB such as energy-efficient lighting systems have been implemented in Nairobi (Mulei, 2020). The barriers to the adoption of GBs in Kenya include lack of institutional support, lack of regulatory and policy tools, socio-economic factors, and inadequate technical and awareness interventions (Khaemba, 2013). The concept is relatively new and its performance information is scanty (Asuza et al., 2022). Therefore, this chapter aimed at determining the extent of the adoption of GB practices in Kakamega municipality in Western Kenya. This municipality is found in one of the largest counties in Kenya in terms of population and therefore, the adoption of GB concepts will help in addressing sustainability problems bedeviling the building construction sector in the country.

2. Methodology

2.1 Study Area

The study was carried out in Kakamega municipality. The municipality is the capital of Kakamega County. Kakamega County is located in the Western part of Kenya and borders Vihiga County to the South, Siaya County to the West, Bungoma and Trans Nzoia Counties to the North, and Nandi and Uasin Gishu Counties to the East (see Fig. 2). The County covers an area of 3,051.3 km^2 and is the second most populous county after Nairobi and the most populous rural county in Kenya (KNBS, 2019). The County comprises 12 subcounties, 60 wards, 187 village units, and 400 community areas.

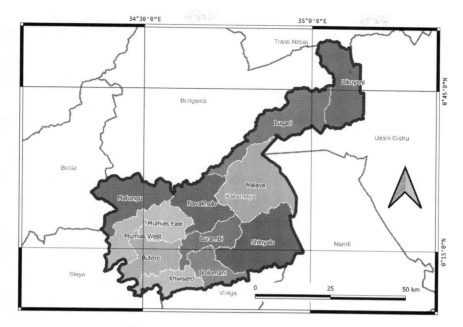

Fig. 2.　Map of Kakamega County.

2.2 Research Design, Data Collection, and Analysis

This study looked into the extent of adoption of GB concepts in commercial and residential buildings using descriptive research design.

The focus was on the commercial and residential buildings within Kakamega municipality. The data were collected through observations, and questionnaires, given to those responsible for the planning, building, and the occupants of the buildings. The degree of adoption was determined using a quantitative method using factors that characterize GBs with a range of variables from minimal to full incorporation of the elements. The factors considered were water efficiency, energy efficiency, sustainable materials, sustainable site practices, and IEQ based on the LEED criteria. The degree of adoption was determined using the Likert scale.

The close similarity in the architecture of these structures and their shared geographic vicinity reduced sampling error.

The population in this study contained commercial and residential buildings in the Kakamega Municipality. The municipality has over 500 buildings. The survey was conducted on 49 buildings both residential and commercial. The survey also involved 15 construction stakeholders which included engineers, quantity surveyors, architects, contractors, and consultants.

The purposive sampling method was adopted for selecting construction stakeholders in this study. This facilitated obtaining data from eligible participants who were willing to offer the needed information. The respondents were classified into engineers, architects, site supervisors, and quantity surveyors.

The non-probability sampling technique was utilized to acquire a representative sample. It is an appropriate method when a completely random sampling method cannot be used to select respondents from the whole population. The respondents can rather be selected based on their willingness to partake in the research (Wilkins, 2011). Thus, a snowball sampling method was used in this study to obtain a valid and effective overall sample size. The snowball method was also used in previous construction management studies (Mao et al., 2015; Zhang et al., 2011), and it allows the gathering and sharing of information and respondents through referral or social networks.

Using this approach, a total of 90 survey questionnaires were administered to collect responses from engineers, consultants, contractors, and caretakers/ occupants of the buildings. Finally, 64 sets of questionnaires with valid responses were returned, yielding a 71.1% response rate.

The study used semi-structured interviews, structured questionnaires, and an inspection checklist where primary data were collected from the field. Semi-structured interviews were conducted by asking open-ended questions to get insights into the various GB aspects in the construction sector in the County. These questions helped in reinforcing the responses from the questionnaires. Questionnaires sent to architects, engineers, quantity surveyors, and property managers provided the majority of the data. Along with inspection observations, additional primary data were gathered from semi-structured interviews with decision-makers in the building and construction industry.

The data were analyzed using descriptive statistics with the help of SPSS version 22 and data were presented using frequency tables and charts.

2.3 Reliability and Validity

Before the main survey, a pilot study was adopted to test the comprehensiveness and relevance of the questionnaire (Li et al., 2011) where 20 questionnaires were administered to respondents who were not part of the main survey. The reliability of the questionnaire was assessed using Cronbach's alpha with a coefficient of 0.7 as the acceptability criterion. The questionnaire was finalized based on feedback from the pilot study. The validity of the questionnaire was assessed using expert interviews and an inspection checklist to verify the quality of the data. Additionally, by restricting the sample population to a specific geographic area (Kakamega municipality) and focusing only on commercial and residential structures homogeneity and the level of data reliability were increased.

2.4 Ethical Considerations

Ethical consideration protocols were observed throughout the study. To obtain the participants' consent, a cover letter outlining the study's objectives was used. They were also given the assurance that the data would be used strictly for scholarly purposes. Although the study intended to include photos, the caretakers of the buildings refused to allow the use of cameras on their property, so no photos were included in the study. The caretakers of the buildings were encouraged to participate voluntarily and were assured that the observations would only be used for the study.

3. Results and Discussion

3.1 Demographics

The responses were received from 15 professionals, (Table 1) i.e., engineers, consultants, quantity surveyors, architects, project managers, and contractors. The majority (80%) of the professionals had an experience of fewer than 5 years while 6.7% each had 6–10 years, 11–15 years, and over 20 years.

In terms of occupancy of the buildings (residential and commercial), the majority were rental (55%) followed by commercial at 31% and owner-occupied at 14% (Fig. 3). This is very important as it affects the green operations in the building.

3.2 Awareness of Green Building Concepts

The professionals were asked about their awareness of GB concepts. All the professionals were aware of energy efficiency and IEQ as part of GB elements (Fig. 4). 93% were aware of water efficiency and sustainable site practices while 80% mentioned the use of sustainable construction materials. This implies that the professionals were aware of GB practices.

3.3 The Extent of Adoption of Green Building Concepts

The adoption of various GB concepts varied with sustainable site practices adopted moderately by more than half (53.3%) of the respondents (Table 2). Sustainable construction materials and energy efficiency were the least adopted with 53.3% either adopting to a little extent or not adopting at all. This correlates with the level of awareness where sustainable construction materials received the

Table 1. Respondents' Demographics.

Profession	Frequency (%)
Architect	13.3
Consultants	13.3
Contractors	13.3
Quantity surveyor	6.7
Engineer	53.3

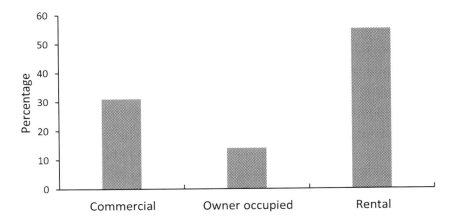

Fig. 3.   Type of Building Use.

lowest rate at 80% (Fig. 4) but surprisingly 100% were aware of energy efficiency as GB technology. Sustainable construction materials require research and validations at different levels by users and this could explain the lower adoption rates. In Hong Kong, sustainable construction materials were rated highly by respondents due to perceived low operational costs (Mesthrige & Kwong, 2018). In general, although the level of awareness of GB technologies in the present study is above 80%, none of the practices was adopted widely by the professionals with an average of only 13.3% either adopting to a great or greatest extent. In a study in Nairobi, Khaemba and Mutsune (2014) found that energy efficiency, water efficiency, IEQ, materials and resources, and sustainable sites were ranked first to fifth, respectively. In Lamu, the coastal part of Kenya, water control strategies

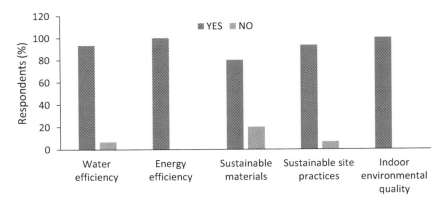

Fig. 4.   Level of Awareness of Green Building Concepts.

Table 2. Adoption of Green Building Concepts.

Green Building Concept	The Extent of Adoption (%)				
	Greatest Extent	Great Extent	Moderate Extent	Little Extent	Not All
Water efficiency	20	6.7	33.3	26.7	13.3
Energy efficiency	13.3	13.3	20	40	13.3
Sustainable site practices	13.3	13.3	53.3	6.7	13.3
Sustainable construction materials	6.7	20	20	33.3	20
Indoor environmental quality	6.7	20	33.3	26.7	13.3

were the least adopted and eco-friendly building and design and energy efficiency were the most adopted aspects (Fadhil, 2015).

Water efficiency is incorporated at least moderately by 60% of the respondents (Table 2). Rainwater harvesting is the most adopted water use efficiency technique with 53.3% at least adopting it greatly (Table 3). Wastewater reuse is the least adopted technique with 53.4% either adopting to a little extent or not adopting it at all. Wastewater recycling is practiced moderately by 40% of the respondents with a third either adopting to a little extent or not adopting at all. The adoption of wastewater reuse and recycling practices is heavily reliant on social and cultural perceptions. The majority of the respondents had moderate use

Table 3. Extent of Adoption of Water Efficiency Techniques.

Green Building Concept	The Extent of Adoption (%)				
	Greatest Extent	Great Extent	Moderate Extent	Little Extent	Not All
Rainwater harvesting	33.3	20.0	20.0	6.7	20
Water use minimization	0.0	13.3	33.3	13.3	40.0
Wastewater recycling	20.0	6.7	40.0	26.7	6.7
Water-efficient fixtures and fittings	13.3	20.0	33.3	26.7	6.7
Water supply submeters	6.7	20.0	46.7	13.3	13.3
Wastewater re-use	6.7	20.0	20.0	26.7	26.7

of water supply sub-meters (46.7%) or not using municipal water for flushing toilets (40%). Reducing wastage through a minimal amount of water use is the least adopted with 40% of the respondents not adopting at all. This, therefore, means that other measures of reducing household water consumption such as demand management tools like water tariffs need to be considered. According to Xie et al. (2017), water-saving habits require personal sacrifices and thus are the least adopted practice.

Landscaping and drainage are important features of GB which ensure the conservation of soil and esthetics around the building. Low impact development (LID) design allows for infiltration and evapotranspiration and thus minimizes storm runoff. 42.9% of the respondents incorporated LID into the site design. 16.7% of those who adopted LID used vegetated storm drains (swales), 50% used bioretention/rain gardens, and 33.3% used permeable pavements. According to Davis (2005), LIDs aim at incorporating as little impervious surface as possible while keeping any runoff on-site as long as possible using natural approaches. Porous pavements may be limited by concerns of groundwater contamination (Dietz, 2007). The drawbacks or barriers to adopting LID are due to the higher implementation and operational costs of conventional stormwater management practices (Davis, 2005).

Integrated solid waste management through reuse and recycling can be effective if the site location and layout provide for efficient storage and collection of segregated solid wastes. In the present study, the majority of the respondents (53.3%) provided dedicated area/areas for the collection and storage of waste materials such as cardboard, plastic, glass, and metals for ease of recycling or reuse programs. As noted by Komolafe et al. (2016), the adoption of recycling programs is affected by policy frameworks or legislation.

Although adoption of sustainable construction materials is low (Table 2), the respondents identified building elements that contain recyclable materials where steel and brick/masonry were the highest (40%) and second highest (26.67%), respectively (Fig. 5). Indeed two-thirds of the respondents said that the buildings

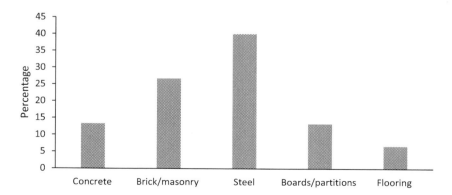

Fig. 5.   Recyclable Building Materials.

promoted the use of recycled materials. All the professionals sampled in the study promoted the use of locally available materials as a sustainability measure. This could be attributed to low cost. 93.3% of the building designers chose construction materials and environmentally friendly construction methods. Promoting the use of recycled materials is important as it reduces the amount of construction and demolition waste generated.

Energy efficiency is important in GB due to the high costs of electricity. The majority (42%) of the residents/users of commercial buildings reported paying monthly electricity bills of between USD 2 and USD 5 followed by 24% who paid between USD 5 and USD 10 (Fig. 6). 16% and 18% paid less than USD 2 and over USD 10, respectively. The occupants/users of commercial/residential buildings were asked to rate how well they think the building was performing in terms of energy efficiency. The majority of the respondents (77.6%) rated the building to be fair in terms of energy efficiency (Fig. 7).

Energy efficiency in a building is a function of lighting, cooling, heating, ventilation, and air conditioning systems. 67% and 69% of the respondents used natural passive systems for heating and cooling, respectively (Fig. 8). Only 29% and 21% used electrical systems for heating and cooling respectively. The findings of the present study were consistent with those found by Komolafe et al. (2016) where a natural cooling system was the preferred option.

The occupants of the commercial/residential buildings were asked about some operational building use concepts relating to energy and water efficiency. The majority (53%) used energy efficiency appliances while 47% used renewable energy sources to minimize electricity bills (Table 4). Strategies for reducing energy consumption and subsequently electricity costs were used by the occupants/users of the buildings where a majority (50%) turned off lights and

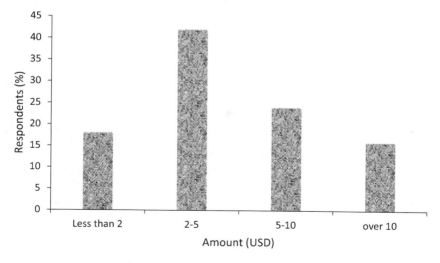

Fig. 6.   Electricity Monthly Cost.

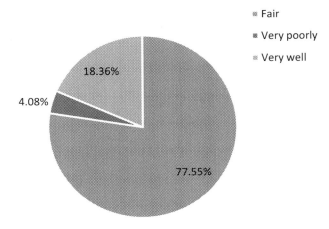

Fig. 7.   Energy Efficiency in Commercial/Residential Buildings.

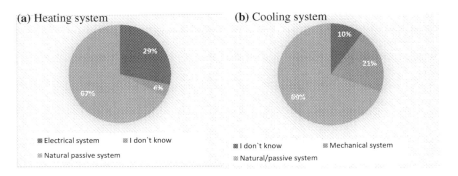

Fig. 8.   Heating and Cooling System.

switched off appliances not in use while 37.5% used energy-saving bulbs (Fig. 9) while only 12.5% used solar systems as an alternative to electrical energy. The results of the present study differed from those found by Fadhil (2015) in Lamu County in Kenya where energy-saving bulbs and the use of solar energy were the most adopted energy efficiency strategies. In Nigeria, Komolafe et al. (2016) found that the installation of solar energy was the least adopted while the use of low-energy bulbs was the most popular energy-saving technique. The majority (55%) of the houses did not use heating, ventilation, and air condition systems. Daily water consumption varied with the majority using between 21–40 liters (37%) and 41–60 liters (27%) (Fig. 10). 18% consume between 61 and 80 liters while only 4% use over 80 liters. Water efficiency was promoted through the

Table 4. Energy and Water Efficiency Measures.

GB Aspect	Adoption (%)		
	Yes	No	Don't Know
Provision for heating, ventilation, and air conditioning	37	55	8
Use or allowance for renewable (solar) energy technologies	47	47	6
Use of energy-efficient appliances and equipment	53	29	18
Use of high efficiency and innovative plumbing fixtures	55	43	2
Use of individual metering in multifamily units	49	51	-
Use of rainwater harvesting	69	31	-
Landscaping with native plants with low water requirement	35	55	10
Use of innovative wastewater technologies, i.e., wastewater treatment plant incorporated into the building to reduce the generation of wastewater and potable water demand	20	78	2

adoption of plumbing fixtures which reduces water wastage such as low water use toilets and efficient water taps (Table 4). Reduction of wastage through individual metering was used by 49% of the respondents in multiple dwelling units. Rainwater harvesting to augment municipal water supply was adopted by 69% of the respondents.

Allowing for natural daylight as much as possible depends on the building design envelopes and the duration of sunlight. The majority (78%) of the

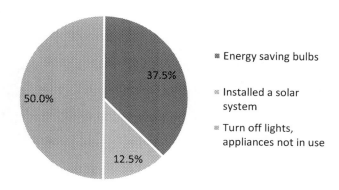

Fig. 9.   Strategies for Minimizing Energy Use and Costs.

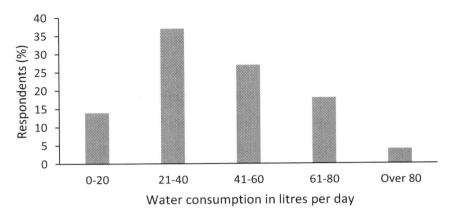

Fig. 10.   Per Capita Daily Water Consumption.

respondents were of the view that the natural daylight exposure is adequate and also the amount of sunlight entering the house is adequate (Fig. 11). This implies that the use of natural light is possible and therefore minimizes the use of artificial lighting during daytime.

IEQ is important in ensuring comfort and preventing sick building syndrome. The ambient environment has both physical and emotional effects on man and is, therefore, of central importance in building design and development (Rabah, 2004). Respondents were asked to rate the quality of air inside their buildings. The majority (96%) reported good quality of air was received in the building. The natural ventilation system was used by 96% of the occupants. This was consistent

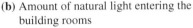

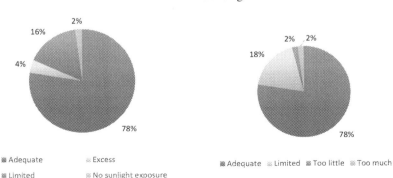

Fig. 11.   Natural Daylight.

with findings by Mesthrige and Kwong (2018) where natural ventilation was ranked as the most suitable GB design feature for Hong Kong. In the present study, 57% of the respondents were of the view that the buildings had provision for better indoor air quality while 41% responded negatively and the rest were indifferent.

The supply of fresh air in a building improves the overall comfort of the occupants. 69% of the respondents were of the view that the overall sense of well-being at the place of residence was fair, 27% very good, and the rest poor. 65.3% of the respondents were moderately comfortable at a place of residence as a result of good indoor air quality (Fig. 12) while 28.6% who were very comfortable inside the house was not stressful at all and the rest were uncomfortable. The adoption rate of GB is premised on the living quality of the occupants. Therefore, the design and application of GB are strongly associated with improvements in IEQ parameters such as indoor air quality, temperature control, and day lighting (Mesthrige & Kwong, 2018).

4. Conclusion

This chapter determined the extent of the adoption of GB concepts in Kakamega municipality. The study used five elements to define the GB, namely, sustainable site practices, sustainable construction materials, energy efficiency, water efficiency, and indoor environmental quality.

From the findings, the following conclusions were drawn.

(1) At least 80% of the professionals were aware of GB concepts with energy efficiency and IEQ having 100% awareness levels. This implies that IEQ and energy efficiency are critical features to be adopted for GBs in Kenya.

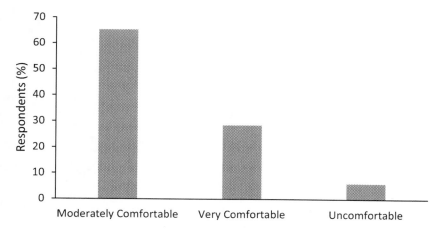

Fig. 12.   Stress Levels Due to Indoor Air Quality.

(2) The use of local construction materials was found to be widely adopted and thus ranked as one of the main drivers of GB adoption. This was due to the low cost of these materials.

(3) Majority of the professionals promoted the use of recycled materials. This is a plus since they minimize the disposal of construction and demolition waste.

(4) Rainwater harvesting was widely adopted in residential and commercial buildings

This study was limited by the small sample size and thus further research is required using a larger sample.

This study considered the extent of adoption of GB concepts only, and therefore research is needed to determine the drivers and barriers to the adoption of these concepts in the Kenyan construction industry. This will help in developing interventions required to facilitate faster adoption.

References

Ahn, Y. H., Pearce, A. R., Wang, Y., & Wang, G. (2013). Drivers and barriers of sustainable design and construction: The perception of green building experience. *International Journal of Sustainable Building Technology and Urban Development, 4*(1), 35–45.

Asuza, D., Diang'a, S., & Mugwima, B. (2022). A maintenance framework for sustainability of green buildings: A case of Nairobi County, Kenya. *East African Journal of Environment and Natural Resources, 5*(1), 127–133.

Bornais, C. (2012). Exploring the diversity of green buildings. *Journal of Green Building, 7*(3), 49–64.

Chan, A. P. C., Darko, A., Olanipekun, A. O., & Ameyaw, E. E. (2018). Critical barriers to green building technologies adoption in developing countries: The case of Ghana. *Journal of Cleaner Production, 172*, 1067–1079.

Choi, C. (2009). Removing market barriers to green development: Principles and action projects to promote widespread adoption of green development practices. *Journal of Sustainable Real Estate, 1*(1), 107–138.

Dahiru, D., Dania, A., & Adejoh, A. (2014). An investigation into the prospects of green building practice in Nigeria. *Journal of Sustainable Development, 7*(6), 158.

Davis, A. P. (2005). Green engineering principles promote low-impact development. *Environmental Science and Technology, 39*, 338A–344A.

Dietz, M. E. (2007). Low impact development practices: A review of current research and recommendations for future directions. *Water, Air, and Soil Pollution, 186*(1), 351–363.

Fadhil, A. (2015). *Adoption of green practices in the hospitality and tourism industry in Lamu County, Kenya.* MA Thesis, University of Nairobi.

Hwang, B. G., & Tan, J. S. (2012). Green building project management: Obstacles and solutions for sustainable development. *Sustainable Development, 20*(5), 335–349.

Khaemba, P. (2013). *Adoption of green building practices and rating system in Kenya: Potentials and barriers.* PhD Thesis, North Carolina Agricultural and Technical State University.

Khaemba, P., & Mutsune, T. (2014). Potential for green building adoption: Evidence from Kenya. *Global Journal of Business Research, 8*(3), 69–76.

KNBS. (2019). *2019 Kenya population and housing census.* Kenya National Bureau of Statistics.

Komolafe, M. O., Oyewole, M. O., & Kolawole, J. T. (2016). The extent of incorporation of green features in office properties in Lagos, Nigeria. *Smart and Sustainable Built Environment, 5*(3), 232–260.

Li, Y. Y., Chen, P.-H., Chew, D. A. S., Teo, C. C., & Ding, R. G. (2011). Critical project management factors of AEC firms for delivering green building projects in Singapore. *Journal of Construction Engineering and Management, 137*(12), 1153.

Mao, C., Shen, Q., Pan, W., & Ye, K. (2015). Major barriers to off-site construction: The developer's perspective in China. *Journal of Management in Engineering, 31*(3), 04014043.

Mesthrige, J. W., & Kwong, H. Y. (2018). Criteria and barriers for the application of green building features in Hong Kong. *Smart and Sustainable Built Environment, 7*(3/4), 251–276.

Mulei, D. N. (2020). *Evaluation of energy efficiency, indoor air quality and sustainability testing of green buildings in Nairobi, Kenya.* MSc Thesis, Jomo Kenyatta University of Agriculture and Technology.

Nguyen, H.-T., Skitmore, M., Gray, M., Zhang, X., & Olanipekun, A. O. (2017). Will green building development take off? An exploratory study of barriers to green building in Vietnam. *Resources, Conservation and Recycling, 127*, 8–20.

Olanipekun, A. O., Xia, B., Hon, C., & Hu, Y. (2017). Project owners' motivation for delivering green building projects. *Journal of Construction Engineering and Management-ASCE, 143*(9).

Olubunmi, O. A., Xia, P. B., & Skitmore, M. (2016). Green building incentives: A review. *Renewable and Sustainable Energy Reviews, 59*, 1611–1621.

Pitts, J., & Jackson, T. O. (2008). Green buildings: Valuation issues and perspectives. *The Appraisal Journal, 76*(2), 115–118.

Rabah, K. (2004). Self-reliant climate-specific energy-efficient solar-powered home in Garissa Kenya. *Architectural Science Review, 47*(1), 9–17.

Robichaud, L. B., & Anantatmula, V. S. (2011). Greening project management practices for sustainable construction. *Journal of Management in Engineering, 27*(1), 48–57.

Wilkins, J. R. (2011). Construction workers' perceptions of health and safety training programmes. *Construction Management & Economics, 29*(10), 1017–1026.

Xie, X., Lu, Y., & Gou, Z. (2017). Green building pro-environment behaviours: Are green users also green buyers? *Sustainability, 9*(10).

Zhang, X., Shen, L., & Wu, Y. (2011). Green strategy for gaining competitive advantage in housing development: A China study. *Journal of Cleaner Production, 19*(2–3), 157–167.

Zuo, J., & Zhao, Z.-Y. (2014). Green building research–Current status and future agenda: A review. *Renewable and Sustainable Energy Reviews, 30*, 271–281.

Chapter 9

Transient Thermodynamic Modeling of Heat Recovery From a Compressed Air Energy Storage System

Mehdi Ebrahimi, David S-K. Ting and Rupp Carriveau

Abstract

Sustainable development calls for a larger share of intermittent renewable energy. To mitigate this intermittency, Compressed Air Energy Storage (CAES) technology was introduced. This technology can be made more sustainable by recovering the heat of the compression phase and reusing it during the discharge phase, resulting in an adiabatic CAES without the need for burning of fossil fuels. The key process parameters of CAES are temperature, pressure ratios, and the mass flow rates of air and thermal fluids. The variation in these parameters during the charge and discharge phases significantly influences the performance of CAES plants. In this chapter, the transient thermodynamic behavior of the system under various operating conditions is analyzed and the impact of heat recovery on the discharge phase energy efficiency, power generation, and CO_2 emissions is studied. Simulations are carried out over the air pressure range from 2,500 to 7,000 kPa for a 65 MW system over a five-hour discharge duration. It is also assumed that the heat loss in the air storage and the hot thermal fluid tank is insignificant and standby duration does not impact the status of the system. This result shows that the system exergy and the generated power are more sensitive to pressure change at higher pressures. This work also reveals that every 10°C increase on the temperature of the stored air can lead to a 0.83% improvement in the energy efficiency. The result of the transient thermodynamic model is used to estimate the reduction in CO_2 emissions in CAES systems. According to the obtained result, a 65 MW ACAES plant can reduce about 17,794 tons of CO_2 emission per year compared to a traditional CAES system with the same capacity.

Pragmatic Engineering and Lifestyle, 171–188
Copyright © 2023 Mehdi Ebrahimi, David S-K. Ting and Rupp Carriveau
Published under exclusive licence by Emerald Publishing Limited
doi:10.1108/978-1-80262-997-220231009

Keywords: Air energy storage; transient thermodynamic analysis; CAES; greenhous gas emission; heat recovery; adiabatic compressed air energy storage

1. Introduction

Global population growth, economic development, and energy structure transformation accelerated the problem of the growing gap between peak and valley of electrical power demand. To overcome this challenge and to efficiently use the surplus electrical power at low-demand times, different types of energy storage systems have been developed (Akinyele & Rayudu, 2014; Ebrahimi, Carriveau, Ting, McGillas, & Young, 2019). Energy storage technologies can improve the efficiency of the electrical grid, increase the power transmission line capability, and improve the power reliability and quality. Among various energy storage technologies, compressed air energy storage (CAES) is one of the large-scale concepts that has the advantages of low cost, high safety factor, and relatively fast response time in a scale of a couple of minutes to reach to the full capacity. However, the waste heat of the compressed air during the charging phase and burning the fossil fuel in the discharge phase to heat up air is an environmental issue in traditional CAES technology (ones with fossil fuel burners). Moreover, to heat up compressed air before entering the expanders, external heat is required and this heat is generally made possible via combustion, resulting in CO_2 emissions. Considering that the traditional CAES system has an adverse impact on the environment, an adiabatic compressed air energy storage (ACAES) system has recently been proposed which recovers compression phase heat to reduce both greenhouse gas (GHG) emissions and thermal pollution (Hartmann et al., 2012; Liu, Li, et al., 2014).

Mohammadi et al. (2017) studied the adverse impact of the traditional CAES systems on the environment. They used an organic Rankine cycle to increase the power production and the cooling capacity by recovering the dissipated heat. In their model, they utilized toluene as the working fluid and reported a roundtrip energy efficiency of 53.9%. Jubeh and Najjar (2012) discussed the recommendations and guidelines for designing expanders in ACAES systems. Zhao et al. (2015) used a high-temperature Kalina cycle system to recover the waste heat of the CAES gas turbine. This improvement increased the round trip energy efficiency by around 8.8% compared to a stand-alone CAES system. Meng et al. (2018) applied a recuperator-equipped organic Rakine cycle to recover the heat dissipated in the expansion and compression cycles. They studied different working fluids and reported that R123 improves the roundtrip efficiency by around 6.7%. Raju and Khaitan (2012) modeled the behavior of air streams by using mass and energy conservation equations. They proposed a method for the calculation of the heat transfer coefficient in ACAES plants. They calibrated the model with the real data from the Huntorf CAES plant with the capacity of 226 MW. Proczka et al. (2013) worked on a design method for the air storage accumulator and identified the size determination criteria. Razmi et al. (2019)

utilized a desalination cycle to restore the waste heat of the turbine and compressors. With this technique, they were able to generate about 80 MW of power for peak shaving as well as production of 38 kg/s and 62.5 kg/s of freshwaters during the off-peak and peak times, respectively. Zhang et al. (2013) studied the thermodynamic effects of air storage accumulator modes on the ACAES system. They reported that different air storage accumulator modes result in the different characteristics of the charge and discharge cycles, thereby affecting the efficiency of the entire system. Zeynalian et al. (2020) used a hybrid unit by combining a traditional CAES system with organic Rankine cycle and CO_2 capture units. With this method, they were able to improve the environmental feasibility of the traditional CAES by removing about 88% of the produced CO_2. Liu, Liu, et al. (2014) performed a thermodynamic analysis of a traditional CAES system. They showed that the roundtrip energy efficiency of their CAES system could be improved by about 10% if a combined cycle was used. They showed that the recovered heat from compressors' intercoolers can significantly improve system efficiency by keeping the steam part of the system on the hot standby mode.

From the review of the literature herein, it can be observed that studies investigating the effect of heat recovery on the performance and emission reduction of the traditional CAES system are not well studied mainly due to the limited number of under-operation CAES plants. Among the few studies that are on the topic, they are mostly limited to the steady-state analysis. Although steady-state analysis is a powerful tool for the analysis of traditional and ACAES systems, it is not suitable for analyzing cycles with dynamic behaviors and large heat capacities. In such cases, transient analysis is necessary to understand the effect of heat recovery in the compression phase and use it in the expansion phase. This is a key field of study that can expedite the process of moving from traditional CAES plants to emission-free CAES plants. According to the authors' best knowledge, globally there is only one under-operation emission-free CAES plant which is a relatively small one with the capacity of about three MWh in Goderich, Canada (hydrostor, 2021). This chapter is thus one of the first attempts to fill this gap by analyzing the transient behavior of ACAES for various operating conditions. The exergy and the energy efficiency of the system are also studied for different initial system modes. It is hoped that the outcomes of this chapter can contribute to a better understanding of the transient behavior of ACAES systems and can be applied for the improvement of the engineering design of ACAES systems and other emission-free CAES technologies.

2. Technology Description

A schematic flow diagram of an ACAES system is shown in Fig. 1. In the ACAES technology, the redundant power in electricity off-peak periods drives the multistage compressor unit to compress air to a state of high temperature and pressure. The compressed air then is stored in an air storage tank after its compression heat is recovered through the thermal subsystem. In ACAES systems the heat recovery eliminates fossil fuel burners from the system thus

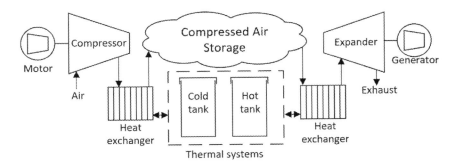

Fig. 1. Process Flow Diagram for ACAES Concept.

resulting in emission-free plants. This modification can also significantly improve the efficiency of the systems. Moreover, the heat recovery reduces the hot air temperature inside the system which leads to removal or a smaller cooling subsystem. The energy release process starts at the electricity demand peak times when the pressurized air from the air storage returns to the system and air is reheated in the heat exchanger. The heated air then enters the turbine and generates electricity.

3. Governing Equations of the Mathematical Model

Applying the conservation law of mass and energy, ideal gas law, and thermodynamic law, the mathematical models of the component of ACAES systems can be established. In this work, the thermodynamic cycle method is applied for the transient thermodynamic modeling of the understudied ACAES system. In literature, it has been shown that this method provides a high accuracy model for the analysis of CAES systems (Fallah et al., 2016) since the thermodynamic specifications of the model are well understood (Ebrahimi, Ting, Carriveau, & McGillis, 2020). Moreover, it was assumed that the temperature of the hot water in the hot tank is constant during the discharge phase and no chemical reaction exists in the system. Constraints of the models were also selected based on our previous works (Ebrahimi, Carriveau, Ting, & McGillis, 2019, Ebrahimi, Ting, Carriveau, McGillis, & Young, 2020).

It should be noted that the temperature drop of the thermal fluid in the hot tank can significantly impact the performance of the CAES systems. There are many parameters that can influence the temperature of the thermal fluid in the hot tank such as the standby duration, insulation material and thickness, thermal fluid type, hot tank size and shape, etc. In real plants, according to the specification of the project, this assumption needs to be checked.

The performance of a CAES system is determined primarily by the mass flow rates of the system fluids, pressure ratios of the compressors and the expanders, the effectiveness of the heat exchangers, and the ambient condition. In this work, we assumed that air behaves as an ideal gas, and the kinetic and potential energies of fluid streams were negligible. To better understand the behavior of the system in the transient phase, it is also assumed that the isentropic efficiency of the turbine and compressor during the discharge phase is fixed (Carriveau et al., 2019). In real-world problems the actual efficiency map of these machines should be used to determine the thermodynamic status of the system at each time step. A logarithmic polynomial function (Lanzafame & Messina, 2000) was used for calculating the specific heats of fluid streams as follows:

$$c_p(T) = a + bT + cT^2 + dT^3 + eT^4 + fT^5 \tag{1}$$

where a, b, c, d, e, and f are the polynomial coefficients and T is the temperature in Kelvin. Considering that the outlet air from the cavern is humid, the enthalpy of humid air at any temperature then can be calculated as follows:

$$h = c_{p-\text{air}}(T) + \left(2500 + c_{p-\text{h}_2\text{o}}(T)\right) \tag{2}$$

Thermodynamic simulation of the models has been conducted based on previous works (Ebrahimi et al., 2019) by simulating the different components of the system based on the conservation of mass and energy. The modeling details of the main system components are described in the next section.

3.1 Main System Components

3.1.1 Air Storage

The air storage performance can impact the efficiency of the system. To avoid making our model location or geology specific, we assume that the air storage is adiabatic. In real problems, according to the specification of the air storage (normally a cavern) including the thickness and properties of the wall rocks (to determine the heat transfer coefficient), and standby duration, the heat transfer rate can be calculated. In our transient model, in any time step, the rate of mass change in the air storage can be determined as:

$$\sum \dot{m}_{\text{in}} - \sum \dot{m}_{\text{out}} = \frac{dm}{dt} \tag{3}$$

where $\dot{m}_{\text{in}}$ and $\dot{m}_{\text{out}}$ are the inlet and outlet mass flow rates to and from the air storage, respectively; $\frac{dm}{dt}$ is the mass accumulation in the air storage, γ is the specific heat ratio, and t is the time. Using the first law of thermodynamics, the temperature and pressure at each time step can be calculated by:

$$T_{c,t} = \frac{1}{m_{c,t}\left(\frac{\gamma-1}{\gamma}\right)\left(m_t T_{\text{in},t} - m_t T_{\text{out},t}\right)} \tag{4}$$

$$P_t = X_t h_t + \frac{1}{V_c\left(n_{c,t} C_{P,t} T_{c,t} - m_t h_{\text{in},t} - m_t h_{\text{out},t}\right)} \tag{5}$$

3.1.2 Heat Exchangers

Heat exchangers are integral parts of CAES systems as they bridge the compression and expansion subsystems via the thermal energy storage subsystem. A well-designed thermal management system eliminates the need for external heat and has the potential to improve the performance of cycles by up to about 80% (Elmegaard & Wiebke, 2011).

The efficiency of heat exchangers is usually evaluated in terms of their effectiveness. This parameter is defined as the ratio of the actual heat transfer rate with respect to the theoretical maximum, that is:

$$\eta = \frac{\dot{Q}_{actual}}{\dot{Q}_{max}} \tag{6}$$

By neglecting heat loss to the environment and transient effects such as heating of equipment metals and residence time of fluids in the exchangers, the rate of heat transfer in a heat exchanger can be calculated by:

$$\dot{Q} = \dot{m}_{hot} C_{p,hot} \left(T_{hot,in} - T_{hot,out} \right) = \dot{m}_{cold} C_{p,cold} \left(T_{cold,out} - T_{cold,in} \right) \tag{7}$$

where $\dot{m}$ is the mass flow rate (kg/s), T is the temperature (K) and C_p is the specific heat capacity at constant pressure (J/(K kg)).

The pressure loss in the heat exchanger can also be calculated with an empirical equation (Yang et al., 2014) as follows:

$$P_{loss} = \frac{0.0083\eta}{1 - \eta} P_{in} \tag{8}$$

where P_{in} is the pressure at the inlet of heat exchangers. According to the defined thermodynamic model and specification of the system, the pressure loss can be between two and 15 kPa for a pipeline with the roughness of 0.004 millimeter and the length of 40–300 meter, respectively. In this work, water is selected as the thermal fluid due to its high thermal capacity and low cost. The use of different fluids would result in a significantly higher price per kWh stored. On the negative side, water changes phase and it is difficult to manage it at high temperatures since designers need to keep water in the liquid phase (to keep the system's cost reasonable) by controlling the pressure of the water line. For example, in this work the maximum water temperature in the hot tank is about 220°C which resulted in a relatively high-pressure line of 5 MPa. Therefore, according to the specification of each project and the thermodynamic condition of each plant, a proper thermal fluid should be selected.

3.1.3 Compressor and Expander

To determine the received power to the compressor and the generated power by the expander, the exit temperature of the compression and expansion processes needs to be calculated. Assuming isentropic processes with constant specific heat, the relation between the pressure and temperature at any stage can be determined by:

$$T_f = T_i \left(\frac{P_f}{P_i} \right)^{\frac{\gamma-1}{\gamma}} \tag{9}$$

where i and f are the initial and final states of air, respectively. The actual power generation at any stage can be calculated as follows:

$$\dot{W}_{comp.} = \frac{\left(\frac{\dot{m}RT_i}{\eta_{comp.}\left(\left(\frac{P_f}{P_i}\right)^{\frac{\gamma-1}{\gamma}}\right)} - 1\right)\gamma}{\gamma - 1} \tag{10}$$

$$\dot{W}_{exp.} = \frac{\left(\frac{\dot{m}RT_i}{\eta_{exp.}\left(1 - \left(\frac{P_f}{P_i}\right)^{\frac{\gamma-1}{\gamma}}\right)}\right)\gamma}{\gamma - 1} \tag{11}$$

Then, the final compression and expansion temperatures at any time step can be calculated using:

$$T_{f,comp.,actual} = T_i\left(1 + \frac{1}{\eta_{comp.}\left(\left(\frac{P_f}{P_i}\right)^{\frac{\gamma-1}{\gamma}} - 1\right)}\right) \tag{12}$$

$$T_{exp} = T_i\left(1 - \eta\left(1 - \left(\frac{P_f}{P_i}\right)^{\frac{\gamma-1}{\gamma}}\right)\right) \tag{13}$$

3.2 CAES Efficiency

To compare the transient efficiency of the traditional CAES and ACAES systems and to be able to study the effect of air property changes during the discharge cycle on the system performance, a discharge energy efficiency (DEE) is defined as follows:

$$\text{DEE} = \frac{\dot{E}_1 + \dot{E}_2 + \cdots + \dot{E}_{k\text{-th}}}{\dot{E}_{CO} + \dot{E}_{HW}} \tag{14}$$

where $\dot{E}_{(k)}$ is the energy of the k-th expander, $\dot{E}_{CO}$ air energy at the outlet of the air storage and $\dot{E}_{HW}$ the energy of hot fluid at the inlet of the heat exchangers. Additional information such as the efficiency and loss in the system components needs to be specified for calculating the other thermodynamic parameters. These inputs that we considered for the transient thermodynamic simulation of the system are summarized in Table 1 (Ebrahimi et al., 2019).

4. Results and Discussion

In this work, transient thermodynamic modeling of heat recovery from a ACAES System with the electrical capacity of 65 MW is investigated and the effect of waste heat recovery on the performance of ACAES system are studied for a wide range of operating conditions. The schematic flow diagram of the understudied CAES plant is shown in Fig. 2.

Table 1. The Assumption of Parameters Considered for Calculation of System Status.

Component	Parameters
First stage expander	$\eta_{\text{Isentropic}} = 89.5\%$
First heat exchanger	Air pressure loss = Equ(1.8); up to 15 kPa Water pressure loss $\cong$ 5.5 kPa Water inlet temperature = 220°C Hot water mass flow rate portion = 0.333
Second stage expander	$\eta_{\text{Isentropic}} = 89.5\%$
Second heat exchanger	Air pressure loss = Equ(1.8); up to 13 kPa Water pressure loss $\cong$ 5.5 kPa Water inlet temperature = 220°C Hot water mass flow rate portion = 0.327
Third stage expander	$\eta_{\text{Isentropic}} = 89.5\%$
Third heat exchanger	Air pressure loss = Equ(1.8); up to 10 kPa Water pressure loss $\cong$ 5.5 kPa Water inlet temperature = 220°C Hot water mass flow rate portion = 0.340
Air storage pipeline	Air pressure loss $\cong$ 6.5% of the air storage presure

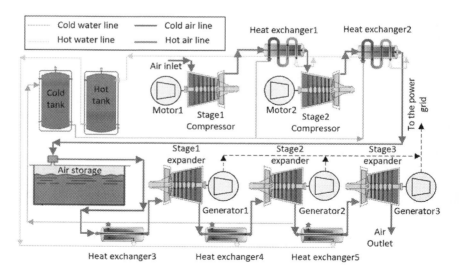

Fig. 2.   Schematic Diagram of the Applied ACAES for the Thermodynamic Analysis.

In the transient modeling, it is assumed that the discharge duration is about five hours whereas the maximum and the minimum pressure in the air storage are 7,000 kPa and 2,500 kPa, respectively. The air and water mass flow rates are considered to be constant at rates of 150 kg/s and 240 kg/s, respectively. It is also assumed that the hot tank is well insulated, and the maximum water temperature drop in the hot tank is less than 2%. It should be noted that when the pressure reaches 2,500 kPa or the system ran out of hot thermal fluid the plant cannot generate any more power. In this condition, it is assumed that the exergy of the plant is zero.

In Table 2 and Table 3 the thermodynamic status of the main component of the system are presented at different time steps for the initial air storage conditions of P = 7,000 kPa and T = 20°C (P7000T20) and P = 6,500 kPa and T = 10°C (P6500T10). From these tables, it can be observed that how the system responds to different initial air storage conditions for two sample state points.

Table 2. The Transient Status of the System at Two Different Time Steps With the Air Storage for Case P7000T20.

State Point	Fluid	Initial Condition			After 2 Hours		
		T(°C)	P(kPa)	Mass (kg/s)	T(°C)	P(kPa)	Mass (kg/s)
Air storage outlet	Air	20.0	7,000	150.0	20.0	5,500	150
Air storage pipeline outlet	Air	12.6	5,895	150.0	12.8	4,579	150
Heat exchanger 1 inlet	Air	12.6	5,886	150.0	12.8	4,570	150
Heat exchanger 1 outlet	Air	210.8	5,878	150.0	213.3	4,562	150
Stage 1 expander inlet	Air	210.7	5,870	150.0	213.2	4,554	150
Stage 1 expander outlet	Air	69.2	1,468	150.0	71.3	1,139	150
Heat exchanger 2 inlet	Air	69.1	1,458	150.0	71.2	1,130	150
Heat exchanger 2 outlet	Air	213.8	1,448	150.0	214.1	1,119	150
Stage 2 expander inlet	Air	213.7	1,436	150.0	214.0	1,107	150
Stage 2 expander outlet	Air	72.5	359	150.0	72.9	277	150

Table 2. *(Continued)*

State Point	Fluid	Initial Condition			After 2 Hours		
		$T(°C)$	P(kPa)	Mass (kg/s)	$T(°C)$	P(kPa)	Mass (kg/s)
Heat exchanger 1 inlet	Air	72.4	343	150.0	72.8	261	150
Heat exchanger 1 outlet	Air	215.3	335	150.0	216.1	253	150
Stage 3 expander inlet	Air	215.1	319	150.0	215.3	236.5	150
Stage 3 expander outlet	Air	78.7	84	150.0	78.8	62.5	150
Hot tank outlet	Water	220.0	5,000	115.0	220.0	5,000	115
High-pressure line	Water	220.0	4,997	38.3	220.0	4,997	38.3
Intermediate pressure line	Water	220.0	4,998	37.6	220.0	4,998	37.6
Low-pressure line	Water	220.0	4,998	39.1	220.0	4,998	39.1
Cold tank inlet	Water	68.2	4,984	115	69.1	4,984	115

Table 3. The Transient Status of the System at Two Different Time Steps With the Air Storage Initial Condition for Case P6500T10.

State Point	Fluid	Initial Condition			After 2 Hours		
		$T(°C)$	P(kPa)	Mass (kg/s)	$T(°C)$	P(kPa)	Mass (kg/s)
Air storage outlet	Air	10.0	6,500	150.0	10.0	5,000	150
Air storage pipeline outlet	Air	2.4	5,417	150.0	2.5	4,112	150
Heat exchanger 1 inlet	Air	2.6	5,408	150.0	2.7	4,102	150
Heat exchanger 1 outlet	Air	211.5	5,400	150.0	214.2	4,094	150
Stage 1 expander inlet	Air	211.4	5,392	150.0	214.1	4,087	150
Stage 1 expander outlet	Air	69.8	1,348	150.0	72.1	1,022	150

Table 3. *(Continued)*

State Point	Fluid	Initial Condition			After 2 Hours		
		T(°C)	*P*(kPa)	Mass (kg/s)	*T*(°C)	*P*(kPa)	Mass (kg/s)
Heat exchanger 2 inlet	Air	69.7	1,339	150.0	72.0	1,013	150
Heat exchanger 2 outlet	Air	214.0	1,329	150.0	214.1	1,003	150
Stage 2 expander inlet	Air	214.2	1,317	150.0	214.3	991	150
Stage 2 expander outlet	Air	72.7	329	150.0	73.0	248	150
Heat exchanger 1 inlet	Air	72.6	313	150.0	72.9	231	150
Heat exchanger 1 outlet	Air	215.3	305	150.0	215.4	223	150
Stage 3 expander inlet	Air	215.1	289	150.0	215.2	207	150
Stage 3 expander outlet	Air	78.8	76	150.0	78.9	55	150
Hot tank outlet	Water	220.0	5,000	115.0	220.0	5,000	115
High-pressure line	Water	220.0	4,997	38.3	220.0	4,997	38.3
Intermediate pressure line	Water	220.0	4,998	37.6	220.0	4,998	37.6
Low-pressure line	Water	220.0	4,998	39.1	220.0	4,998	39.1
Cold tank inlet	Water	65.1	4,984	115	66.0	4,984	115

The air pressure and the total mass of the accumulated air in the storage tank during the discharge process are plotted with respect to time in Fig. 3. The corresponding temperature and the amount of stored hot water in the hot tank are shown in Fig. 4. As discussed, the hot tank is assumed to be well insulated, and the water temperature drop in this condition is around 1.6°C. In real conditions, according to the insulation thickness and the standby duration between the charge and the discharge phases, the decrease in water temperature can be customized.

The exergy of the ACAES system and the total exergy of air are plotted as a function of the elapsed time during the discharge cycle in Figs. 5 and 6, for Case P7000T20 and Case P6500T10. By comparing these two figures it can be observed that the effect of initial temperature compared to the system pressure is insignificant for the considered system operation range.

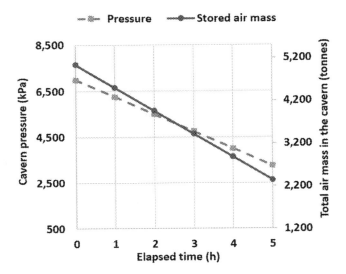

Fig. 3.   Air Pressure and the Total Mass of the Accumulated Air in
the Storage Tank During the Discharge Process.

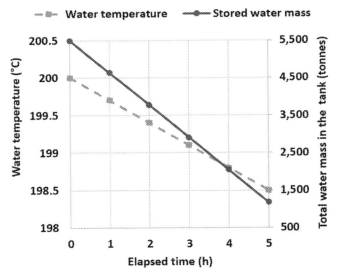

Fig. 4.   Hot Water Temperature and the Mass of the Available Water
in the Hot Tank During the Discharge Process.

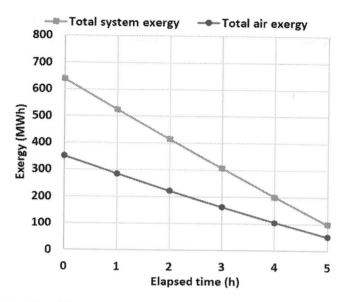

Fig. 5.   Total Exergy of the System and the Air Exergy During the Discharge Process for Case P7000T20.

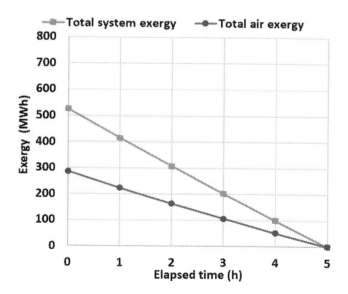

Fig. 6.   Total Exergy of the System and the Air Exergy During the Discharge Process for Case P6500T10.

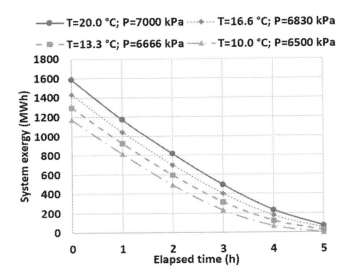

Fig. 7.   Total Exergy of the System During the Discharge Process at
Various Initial Conditions.

In Fig. 7 the effect of initial air storage condition on the exergy of the system is depicted. From this figure, it can be seen that the system exergy and the generated power are more sensitive to system status change at higher pressures and lower air temperatures. On other words, when the discharge cycle starts at lower air temperature, there will be less change in the discharged energy during the expansion phase. Moreover, at higher system pressure, the variation in the discharged energy is larger. For example, in the first discharge hour, the net generated energy is about double of last discharge hour.

In Fig. 8 the transient discharge efficiency of the system for four different initial air storage conditions are compared. This figure shows that for every 10°C increase in cavern air temperature, the efficiency of the system is improved by about 0.8%. As it can be observed, the efficiency of the system is greater at higher air system pressures. However, running the system at higher operating pressure implies higher capital and maintenance costs. Therefore, according to the size of the facility and the designed discharged duration, a cost-benefit analysis should be performed to determine the optimum system operating pressure.

In traditional CAES, air is heated up after each stage of the expansion phase to prevent reaching the freezing temperature. The required energy for heating up the air is supplied from an external energy resource which is usually a fossil fuel burner. In the ACAES system, however, this energy is provided from the recovery of the wasted energy in the charge cycles. In this part, it will be analyzed how

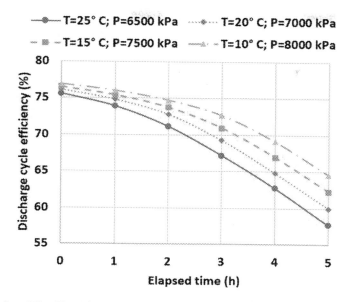

Fig. 8.   The Transient Discharge Efficiency of the System at Various Initial Air Storage Conditions.

much CO_2 emission can be reduced by the construction of ACAES instead of a CAES system for the case P7000T20 and P6500T10. To do such analysis, the equivalent CO_2 emission for heating up that volume of water in the cold tank to 220°C is calculated. Then the reduction in CO_2 dispersion can be determined for the generation of each megawatt-hour of energy (CO_2-e/MWh) for both sample test cases. It should be noted that this process causes various GHG emissions; however, in this study we only focused on the CO_2 emissions.

Considering that the CO_2 emissions in burners vary depending on the burning process and the feedstock, in this part we compared the CO_2 production rate using black and brown coal and natural gas as the most common fossil fuel globally. In all fossil fuels, hydrogen is separated from carbon, producing large amounts of CO_2 emissions. The CO_2 emission roughly is 342–396 kg CO_2-e/MWh for black coal, 367–378 kg CO_2-e/MWh for brown coal, and 198–234 kg CO_2-e/MWh for natural gas (Longden et al., 2022; Muradov, 2017).

The annual CO_2 emissions reduction for Case P7000T20 and Case P6500T10 are presented in Table 4. From this table, it can be observed that by recovering heat and going with an ACAES system, up to 17,794 tons of CO_2 emissions can be prevented for the studied conditions.

Table 4. Annual CO_2 Emissions Reduction by Using the ACAES System Compared to Traditional CAES Systems From the Combustion of Brown Coal, Black Coal, and Natural Gas, for the Heating Up of the Water in the Cold Tank.

Fossil Fuel Type	Tone CO_2 Emission Reduction	
	P7000T20	P6500T10
Black coal	17,622	16,363
Brown coal	17,794	16,523
Natural gas	10,316	9,579

5. Conclusion

CAES technology relies on the compression and expansion of air. As such, thermal management has a significant impact on the performance and environmental friendliness of the system. In this work, transient heat transfer has been studied through the thermodynamic simulation of ACAES plants. The transient thermodynamic modeling was performed for a system with a capacity of 65 MW with a discharge duration of five hours at various operating conditions. The first part of this work showed that both the system exergy and the generated power are more sensitive to pressure change at higher pressures and lower air temperatures. For example, it was shown that in the first discharge hour, the net generated energy was about double of that of the final discharge hour. Also, it was shown that when the discharge cycle started at lower air temperature, there was less change in the discharged energy during the expansion phase.

This work also revealed that the discharge energy efficiency of the system is higher at higher air storage temperature. For example, it was determined that for every 10°C increase in cavern air temperature, the efficiency of the system improved by about 0.8% for the studied operating range. This work showed that ACAES systems are emission-free with higher efficiencies and greater energy generations compared to traditional CAES systems. For instance, it was determined that for a system with a capacity of 65 MW, a ACAES plant can reduce the CO_2 emission up to around 18,000 tones/year in comparison to its CAES counterpart.

References

Akinyele, D. O., & Rayudu, R. K. (2014). Review of energy storage technologies for sustainable power networks. *Sustainable Energy Technologies and Assessments, 8,* 74–91.

Carriveau, R., Ebrahimi, M., Ting, D. S.-K., & McGillis, A. (2019). Transient thermodynamic modeling of an underwater compressed air energy storage plant:

Conventional versus advanced exergy analysis. *Sustainable Energy Technologies and Assessments, 31*, 146–154.

Ebrahimi, M., Carriveau, R., Ting, D. S.-K., McGillas, A., & Young, D. (2019). Transient thermodynamic assessment of the world's first grid connected UWCAES facility by exergy analysis. In *IEEE, BREST*, France.

Ebrahimi, M., Carriveau, R., Ting, D. S.-K., & McGillis, A. (2019). Conventional and advanced exergy analysis of a grid connected underwater compressed air energy storage facility. *Applied Energy, 242*, 1198–1208.

Ebrahimi, M., Ting, D. S.-K., Carriveau, R., & McGillis, A. (2020). Hydrostatically compensated energy storage technology. In *Green energy and infrastructure* (pp. 1–25). Taylor & Francis.

Ebrahimi, M., Ting, D. S.-K., Carriveau, R., McGillis, A., & Young, D. (2020). Optimization of a cavern-based CAES facility with an efficient adaptive genetic algorithm. *Energy Storage, 2*(6), e205.

Elmegaard, B., & Wiebke, B. (2011). Efficiency of compressed air energy storage. In *Proceedings of 24th International Conference on*, Novi Sad, Serbia.

Fallah, M., Mahmoudi, S. M. S., Yari, M., & Akbarpour Ghiasi, R. (2016). Advanced exergy analysis of the Kalina cycle applied for low temperature enhanced geothermal system. *Energy Conversion and Management, 108*, 190–201.

Hartmann, N., Vöhringer, O., Kruck, C., & Eltrop, L. (2012). Simulation and analysis of different adiabatic compressed air energy storage plant configurations. *Applied Energy, 93*, 541–548.

hydrostor. https://hydrostor.ca/. Accessed on January 15, 2021. [Online].

Jubeh, N. M., & Najjar, Y. S. (2012). Green solution for power generation by adoption of adiabatic CAES system. *Applied Thermal Engineering, 44*, 85–89.

Lanzafame, R., & Messina, M. (2000). A new method for the calculation of gases enthalpy. In *35th Intersociety Energy Conversion Engineering Conference and Exhibit*.

Liu, W., Li, Q. Q., Liang, F., Liu, L., Xu, G., & Yang, Y. (2014). Performance analysis of a coal-fired external combustion compressed air energy storage system. *Entropy, 16*, 5935–5953.

Liu, W., Liu, L., Zhou, L., Huang, J., Zhang, Y., Xu, G., & Yang, Y. (2014). Analysis and optimization of a compressed air energy storage-combined cycle system. *Entropy, 16*(6), 3103–3120.

Longden, T., Beck, F. J., Jotzo, F., Andrews, R., & Prasad, M. (2022). 'Clean' hydrogen? – Comparing the emissions and costs of fossil fuel versus renewable electricity based hydroge. *Applied Energy, 306*, 1–14.

Meng, H., Wang, M., Aneke, M., Luo, X., Olumayegun, O., & Liu, X. (2018). Technical performance analysis and economic evaluation of a compressed air energy storage system integrated with an organic Rankine cycle. *Fuel, 211*, 318–330.

Mohammadi, A., Ahmadi, M. H., Bidi, M., Joda, F., Valero, A., & Uson, S. (2017). Exergy analysis of a combined cooling, heating and power system integrated with wind turbine and compressed air energy storage system. *Energy Conversion and Management, 131*, 69–78.

Muradov, N. (2017). Low to near-zero CO2 production of hydrogen from fossil fuels: Status and perspectives. *International Journal of Hydrogen Energy, 42*(20), 14058–14088.

Proczka, J., Muralidharan, K., Villela, D., Simmons, J., & Frantziskonis, G. (2013). Guidelines for the pressure and efficient sizing of pressure vessels for compressed air energy storage. *Energy Conversion and Management, 65,* 597–605.

Raju, M., & Khaitan, S. K. (2012). Modeling and simulation of compressed air storage in caverns: A case study of the Huntorf plant. *Applied Energy, 89*(1), 474–481.

Razmi, A., Soltani, M., Tayefeh, M., Torabi, M., & Dusseault, M. (2019). Thermodynamic analysis of compressed air energy storage (CAES) hybridized with a multi-effect desalination (MED) system. *Energy Conversion and Management, 199,* 1–11.

Yang, K., Zhang, Y., Li, X., & Xu, J. (2014). Theoretical evaluation on the impact of heat exchanger in advanced adiabatic compressed air energy storage system. *Energy Conversion and Management, 86,* 1031–1044.

Zeynalian, M., Hajialirezaei, A. H., RezaRazmi, A., & Torabi, M. (2020). Carbon dioxide capture from compressed air energy storage system. *Applied Thermal Engineering, 178,* 1–10.

Zhang, Y., Yang, K., Li, X., & Xu, J. (2013). The thermodynamic effect of air storage chamber model on advanced adiabatic compressed air energy storage system. *Renewable Energy, 57,* 469–478.

Zhao, P., Wang, J., & Dai, Y. (2015). Thermodynamic analysis of an integrated energy system based on compressed air energy storage (CAES) system and Kalina cycle. *Energy Conversion and Management, 98,* 161–172.

Chapter 10

Trawl Fisheries Management and Conservation in Malacca Straits

Hoong Sang Wong and Chen Chen Yong

Abstract

This chapter provided systematic and comprehensive analysis on trawl fisheries management and conservation measures in the Straits of Malacca. Detailed analysis is conducted on Malaysian fishery management framework particularly domestic country's trawl fishery status, legal structure, input-control strategies, ecosystem protection plan, pollution, law enforcement, and complementary measures that designed to reduce and prevent overfishing in the exclusive economic zone (EEZ) of Malacca Straits. Gaps and challenges found in existing trawl fisheries literature are presented followed by recommendations for improvement in the management and conservation of trawl fisheries.

Keywords: Trawl fisheries; Malacca Straits; legal framework; conservation; overfishing; sustainability

1. Introduction

South East Asia (SEA) region is one of the most productive multispecies fisheries in the world which produces more marine captured fishes than North America, Europe, or Africa (Williams & Staples, 2010). Malacca Straits, located in the center of SEA region, is considered one of the most productive fisheries with estimated marine resources at USD 7 billion, which includes both market and nonmarket resources (Jagerroos, 2016). Malacca Straits alone contributes more than 700,000 tons of marine fish or half of total marine captured fishes in Malaysia over the past decade. Trawlers are the most important commercial vessel in the Straits which have contributed roughly half of the total marine landings even though trawlers only made up about 13.4% of total licensed fishing vessels in 2011 (Nuruddin & Isa, 2013). However, trawl fisheries in the Straits of Malacca, especially those near coastal areas, have been overfished and very few

Pragmatic Engineering and Lifestyle, 189–216

Copyright © 2023 Hoong Sang Wong and Chen Chen Yong

Published under exclusive licence by Emerald Publishing Limited

doi:10.1108/978-1-80262-997-220231010

stocks remained to be underexploited. Total catch, though, has been on the rise but this often masked the subtle change in the composition of fish being caught, with increasingly less of high-value commercial fish, more of small and low-value fish including trash fish. Main causes are attributed to excess capital, destruction, and reduction of bio-rich coastal ecosystems such as mangroves, coral reefs, and sea grass that function as fish nurseries (Perrings, 2016).

Over the last four decades, substantial scholarly work related to fishery management in Malaysia's Malacca Straits had been conducted; however, there is a lack of comprehensive and coordinated research on Malaysia's fishery management at the domestic and regional levels that specifically addresses overfishing, fragile ecosystem, and law enforcement in mainstream literature. This chapter intended to present an up-to-date comprehensive and systematic analysis of Malaysian fishery management framework particularly domestic country's trawl fishery status, legal structure, input-control strategies, ecosystem protection plan, pollution, law enforcement, and complementary measures that designed to reduce and prevent overfishing in the exclusive economic zone (EEZ) of Malacca Straits. Gaps and challenges in the systematic analysis of the country's commercial trawl fleet are presented followed by recommendations for improvement in the management and conservation of trawl fisheries. This chapter employs scholarly written peer-reviewed articles and gray literature like official government or international body reports that provide valuable information for analyzing contemporary management strategies for sustainability.

2. Historical Background of Trawl Fleets in the Straits of Malacca

The length of the Straits of Malacca is about 805 km whereas width varies from 17 km to 322 km, with uneven depths vary from below 10 m to above 70 m (Jagerroos, 2016). Given the southern part is much narrower compared to the northern part of the Straits, coupled with the fact that it is the second busiest commercial shipping lane in the world, and so this explains why most commercial fishery is concentrated in the northern part (from northern Selangor state onwards) of the Straits (WWFMy, 2013). Malaysia shares 11,320 square km of Malacca Straits' EEZ waters with her neighbor, Indonesia.

The marine fishery resources in the Straits of Malacca are blessed with abundance of marine fish species which can be classified into two broad categories: (1) demersal fishes such as marine catfish, threadfin bream, shrimp, red snapper, grouper, shark, and ray; (2) Pelagic species like fringe scale sardine, yellow stripped trevally, Spanish and Indian mackerels, tuna, wolf herring, and anchovies. Peak fishing season in the Straits of Malacca usually occurs during the Southwest Monsoon season starting from May to September characterized by winds that blow from the southwest, the rise of temperature in the Asian continent, and reduced rainfall. Though commercial fishing vessels such as trawlers and purse seines are mainly targeting demersal and pelagic species, respectively, but they are also catching crustacean (prawns and crabs), molluscs (mostly squids and octopuses), and mixed fish (Islam et al., 2011). In 2015, the Department of Fisheries, Malaysia (DOFM, 2021), reported that the landings of marine finfish, crustacean, and molluscs were worth more than RM 2.1 billion (USD 550 million). Total fishery sector and related employment accounted for about 1.3% of GDP in 1990 and fluctuate around it thereafter. In addition, commercial fishery has also created a fairly large number of downstream employments and incomes such as fish trading and processing, boat making, gear and net manufacturing, ice-making, and cold storage (Yahaya & Abdullah, 1993).

Commercial fisheries in Malaysia involve primarily three key types of gear such as trawl nets, the purse seine, and anchovy purse seine while artisanal fisheries include shellfish collection, hooks and lines, and drift gill nets. Marine landings were mostly consisted of demersal and pelagic finfish (82%), and the rest were mainly squid and prawns (18%). Trawlers, mostly in the range of 20–39 tonnages and 40–70 tonnages, accounted for most of the demersal fish landings. Past trawl surveys in the Malacca Straits showed that the northern region seabed is mainly made up of mud which is suitable for trawling but less ideal for trawling in the southern region because of rock and uneven ground (Ahmad et al., 2003). Historically, commercial fishing had been subsistence-based and mainly relied on traditional gears like traps and fishing stakes with a mere harvest of 50,000 tonnages dated back in 1949. However, the increased knowledge of fishing grounds and better fishing technology over time depressed fish stock productivity under an open access environment (Graham, 1935).

In the early 1960s, commercial mechanized trawlers were introduced into the Malacca Straits and thereafter the number of mechanized boats had increased to about 6,000 vessels in 1981 – a surge in the efficiency and productivity of commercial trawl fishing fleet (Butcher, 1996; Kirkley et al., 2003). Commercial trawl fishing is traditionally dominated by Chinese fishermen, mainly Teochew-speaking Chinese immigrants from coastal region of south-western China, whose forefathers are experienced seafarers with long history of marine fishing. Due to China's political turmoils and wars in late nineteenth and early twentieth centuries, these new immigrants migrated and settled down along the coast of Malacca Straits. With hard work, frugality, and acute business mindset, these family-run fishing enterprises gradually invested into modern fishing vessels like trawlers and gained substantial control over trawl fishing businesses in the Malacca Straits over the last six decades. Family-owned trawlers are mainly composed of mini trawlers (<10 tonnages) and medium-sized trawlers (20–39 tonnages and 40–70 tonnages) which usually operating within 30 nautical miles from the shoreline and have been consistently producing more than 80% of total demersal fish catches in the Malacca Straits (Islam et al., 2011; Williams & Staples, 2010).

However, DOFM's trawl field research in 1997 showed that the estimated density in the coastal area of the upper northern part of the Straits had already experienced a drastic drop – about 50% and 11% of 1981–1982 and 1971–1972 trawl surveys, respectively (Ahmad et al., 2003). In terms of mean catch rates, inshore harvest had reduced from 74.5 kg per hour in 1971–1972 to only 22.7 kg per hour in 1991 and then slipped further down to 18.6 kg per hour in 1997. Offshore harvests had dropped from 116.7 kg per hour in 1987 to 33.3 kg per hour in 1997 – a drastic drop of 75% over a ten-year period even though offshore fishing just started in the mid-1980s; however, deepest stratum (92–185 m) only suffered 24% drop over the same period (Ahmad et al., 2003). In 1981, the number of registered trawlers exceeded 5 thousand vessels: 2,725 units of mini trawler (<10 tonnages), followed by 2,342 units of small-medium trawler (10–39.9 tonnages), and 254 medium-sized trawlers (20–69.9 tonnages). However, by 2018, depressed fisheries caused the total number of trawlers to drop by 50%, especially mini trawlers.

Similar drastic drop also occurred in the Indonesian side of Malacca straits, which prompted the neighboring Indonesian government to ban trawling in the late 1970s; however, Malaysia used different approach to deal with the problems like building artificial reefs, establishing marine parks, increasing research and development as well as encouraging offshore fishing (Jagerroos, 2016; Martosubroto, 2002). Martosubroto (2002) also noted that both countries had attempted to resolve the overlapping claim over the Malacca Straits in 1976 and in 1985 but nothing came into fruition.

3. Fisheries Regulation to Promote Sustainability

To promote responsible fishing in the Straits of Malacca, the Malaysian government has drafted a fishery legal framework according to guidelines established by international bodies such as United Nations Convention on the Law of the Sea (UNCLOS). For example, the Third United Nations Convention on the Law of the Sea (UNCLOS) 1982 granted property rights to coastal states to manage marine resources within 200 nautical miles from the coastline (Mohamed, 1991). UNCLOS Article 56 granted coastal states the sovereign rights to explore, exploit, conserve, and manage the natural resources whether living or nonliving in the seabed, subsoil, and super adjacent waters within the EEZ. Coastal states are not only allowed to have greater access to potentially rich marine resources but are also expected to implement a legal framework and execute law enforcement over the EEZ to preserve these resources (Ahmad, 2011). Subsequently, Articles 61 and 62 emphasized two important principles entrenched within the EEZ system of the CLOS: (1) The sovereign rights bestowed on coastal states required that they discharged their duties and obligations to ensure marine resources are utilized and developed in a sustainable manner, (2) Coastal States needed to develop management and conservation strategies based on best scientific evidence to protect fisheries resources (including migratory fish species) from being overly exploited. Article 61(3) also further required coastal states to employ conservation measures designed to maintain or restore populations of harvested species at levels that produce a maximum sustainable yield (MSY).

The above treaty helped to shape Malaysia's Exclusive Economic Zone Act of 1984 and Fisheries Act of 1985. In the subsequent years, other international agreements of UNCLOS, and the Food and Agriculture Organization (FAO) such as the FAO Agreement on Fishing Vessels on the High Seas (1993), the FAO Code of Conduct for Responsible Fisheries (1995), the United Nations

Agreement on Conservation and Management of Straddling Fish and Migratory Fish Stocks (1995), and the FAO International Plan of Actions for the Management of Fishing Capacity (1999) have helped to fine-tune the Fisheries Act of 1985 (amended 1993), and developed the National Plan of Action for the Management of Fishing Capacity 1 2007–2010 (NPOA 1) and NPOA 2 (2015) to prevent and conserve overfishing as well as to identify an optimal fishing capacity (DOFM, 2015; Gopinath & Puvanesuri, 2006). Fig. 1 below provided a timeline for the development, regulation, and management of fisheries resources.

Consequently, two major fisheries agencies were formed, namely, the DOFM and the Fisheries Development Authority of Malaysia (FDAM), and were responsible for drafting and implementing effective management plans and measures in accordance with the rights and obligations that were granted by both UNCLOS and FAO (Ahmad, 2011). The main functions of DOFM were to oversee the development, management, and regulation of the fisheries sector while FDAM's roles were mainly focused on improving the living standard of the fishermen through a variety of programs, such as constructing landing infrastructure, promoting ecotourism, and developing marketing network for fisheries product along with issuing imports and exports licenses. DOFM was the principal federal institution that managed and promoted sustainable marine fisheries, while FDAM aimed to improve the socioeconomic aspects of fishermen (Abdullah, 1995). Over the last four decades, the fishery authority has implemented several management measures such as targeted MSY, input control policies, ecosystem protection strategies, law, and enforcement to achieve optimum fishing effort in the EEZ particularly inshore fisheries.

However, there were some strong criticisms against using the MSY's approach to determine the optimum yield level even though it had the benefit of using only minimal data such as annual catch and effort figures. Critics argued that the single-species MSY approach was based on oversimplified assumptions of the

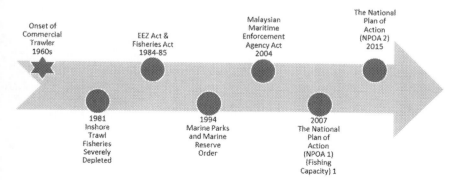

Fig. 1. Important Fisheries Development, Regulation, and Conservation Timeline.

dynamic and complex nature of fisheries and was not suitable for use in rather complex multispecies and multigear fisheries (Haddon, 2011; Hilborn & Walters, 1992). Past research (Kirkley et al., 2003; Wong & Yong, personal communication, 2018) showed that trawlers (including purse seines) targeted not just a single species, but rather, they targeted a variety of high and low commercial marine demersal species, such as grouper, snapper, prawn, and squid. In addition, ambiguous, unclear laws, and regulations plus power struggle between federal and state governments also eroded the effectiveness of management and enforcement measures. For example, state authority was responsible for inland fisheries and aquaculture regulations; however, marine fisheries and aquaculture were under the jurisdiction of the federal authority. The findings of Malaysia's ocean policy project showed that neither the Fisheries Act nor state enactments provide adequate protection for the management of marine and coastal biological diversity including sedentary species and their dependent ecosystems (Saad et al., 2013).

4. Restrictive Input Control Approach

4.1 Strict Rules on Fishing Gears and Mesh Size

To reduce excessive and destructive trawl fishing, Malaysian fishery authority has imposed a ban on the use of dynamite and cyanide fishing practices to avoid causing serious damage to coral reefs, seagrass beds, and benthic organism. Besides, fishery management has also discouraged the use of nonselective fishing gear to reduce significant amount of bycatch and discards over the years. Other destructive gears like pair trawl, push net, and beam trawl were not allowed in local fisheries (Kaur, 2014). In addition, the use of small-coded mesh size, such as 26.6 mm mesh size by prawn trawlers, had resulted in a proportionally high percentage of trash fish landings, roughly four times the weight of young trash fish landed for every unit weight of prawns caught (Blaber et al., 2000). Trash fish including low-value fish and juvenile of high-value species were usually sold to the fish meal industry to be processed into fish pellet for agriculture and fish farming sectors.

To minimize unwanted and accidental large bycatch, the Fisheries Act 1985 specified that Trawler's code end mesh size of trawl nets must be at least 38 mm (1 ½ inches) or greater, but shrimp trawler's code end was smaller and in between of ³⁄₄ − 1 inch, while purse seines' mesh size ranged from 7.8 mm to 100 mm (smaller mesh size for anchovy purse seines operating within 1–8 NM). However, research conducted by Nuruddin and Isa (2013) revealed that many fishermen failed to observe the mesh size regulations over the last three decades due to lax enforcement. Moreover, the existence of many unregistered small trawlers and push netters has also contributed to the overfishing problem. Though there was an attempt to implement the strict mesh size rule in 2006 but the action plan was derailed by a massive demonstration by some local trawlers in certain regions and through political interference. Besides, Basir and Bakar (2011) pointed out that the new mesh size regulation did not provide detailed specification for the design of trawl net which created a loophole for some ingenuine trawl fishermen to modify fishing nets by greatly increasing the vertical opening of the wing mesh size which increased fishing mortality for pelagic species as well. To prevent such loophole, fishery authority now limited the mesh size of the trawl wing and the length of the head rope to a limit of 5 m and 40 m, respectively.

DOFM has also encouraged the skippers of commercial fishing vessels to adopt selective fishing gear (e.g., Juvenile and Trash Excluder Devices (JTED), Malaysian *Acetes* (shrimp) Efficiency Device (MAED), and other selective fishing gear options) to prevent the landing of high amount of bycatch (DOFM, 2015). Broadhurst et al. (2012) research showed that the simplified innovative Nordmore grid and radial escape section prawn nets might reduce the weight of bycatch by as much as 36%. Similarly, some European countries like England, Norway, and Iceland had built "square mesh" panels into the trawl net instead of using the traditional diamond-shaped mesh to prevent the accidental catch of juvenile fish with high commercial value or unwanted species (Saharuddin, 1995). However, the use of "square mesh" or "diamond mesh" to reduce unwanted bycatch is very much correlated to the types of fishes that live in the specific fisheries. To select the "right mesh size and shape," Matsushita and Ali (1997) suggested that further research and development must be carried out to determine species habitat structure and behavior due to seasonal differences and maturity of species at separate fisheries to minimize unwanted catch. To conserve local ecosystem habitats, Saharuddin (1995) recommended the usage of environmentally friendly and cost-saving fishing methods, such as tow-baited lines or bottom-set nets on the seafloor, to prevent trawlers from dragging along the sea floor and inflict heavy damage to local fishery habitats.

Wong and Yong's (personal communication, 2018) field research showed that fishermen often abandoned fish nets, both intentionally or accidently, and this might cause harm to the fishing habitats and marine creatures. Sometimes, fish nets snagged in suboptimal fishing conditions or abandoned by fishers to escape from the detection of maritime enforcement agencies in the open seas. Ghost nets are a worldwide problem, particularly in certain places like the Gulf of Carpentaria where the remote coastlines were littered with up to three tons of derelict nets per kilometer in any given year, mostly originated from neighboring Indonesia (Richardson et al., 2019). Derelict nets could cause gear damage, reduced economic returns, endangered marine lives, and ecological habitats. However, there were no such studies on the Straits of Malacca in contemporary literature. Therefore, more research is needed not only to assess the damage caused by the ghost nets problem in Malaysia but also to help to develop eco-friendly and biodegradable fishing gears to protect marine creatures, improve natural ecosystems, and provide safer passage.

4.2 Prohibitive Zoning System and Licensing Requirements

To further contain overfishing, the Malaysian government had implemented a new zoning system which restricted commercial fishing vessels such as trawlers and purse seines to a predefined fishing zone: artisanal gears only in zone A within 5 NM from shoreline, small fishing vessels (<40 tonnages) in zone B (5–12 NM), medium size fishing vessels (40–70 tonnages) in zone C (12–30 NM), and those big fishing vessels above 70 tonnages were allowed only to operate in zone D (>30 NM) (DOFM, 2015). The other objective of the new fishing zone regime was to reduce increasing conflict between traditional fishermen and commercial trawl operators due to severely depleted inshore marine resources (Alam et al., 2002). To protect coastal fisheries, DOFM also proclaimed a moratorium on trawler license in the inshore region, but at the same time issued new licenses for offshore and deep-sea fishing (Yahaya, 1988) because the fishery authority intended to encourage trawlers to fish in the underexploited offshore fisheries. However, imprecise vessel specifications and asymmetric of information had created opportunity for trawl owners to increase fishing power through engine power upgrade, adoption of modern fishing equipment, modification of fishing gears, and extended fishing expeditions (Alam et al., 2002). In other words, ingenuine fishermen had increased their fishing capabilities which had partly offset the effort reduction plan under the license moratorium and zoning system. Table 1 showed the structural changes in the trawl fishing fleet in Malacca Straits between 1981 and 2018 period which revealed that smaller trawlers were being marginalized by larger trawlers due to depressed fish stock and strict input control measures.

Another issue also arose from the implementation of the above programs, which touched on the equitable sharing of fishing rights between small commercial fishing vessel operators and artisanal fishermen. This property right issue was further complicated by socioeconomic elements because Chinese ethnic group dominated the industrial fisheries while Indigenous Malays were mainly confined to artisanal fisheries. Small trawlers (or mini trawlers) perceived that these new policies denied them the right to catch prawn resources near inshore areas that brought them with lucrative rewards (Pauly & Chua, 1988). As a result, many mini trawlers (below 10 tonnage) suffered substantial financial losses due to the introduction of zoning and restrictive licensing policies, but these new rules also

Table 1. Trawl Gears by Tonnage Classes in Malaysia's Malacca Straits 1981–2018.

Year	Registered Trawlers by Tonnage Classes (t)				Total
	0–9.9 t	10–24.9 t	25–39.9 t	40–69.9 t	
1981	11,124	1,370	972	254	13,720
1990	11,074	2,416	579	411	14,480
2000	9,069	1,749	608	483	11,909
2010	5,724	1,929	652	658	8,963
2018	3,720	986	705	745	6,156

Note: Data of registered trawler tonnage classes in the Malacca straits were taken from the Department of Fisheries, Malaysia (1981–2018).

caused high layoff among hired crews in the industrial fisheries sector, which made up 67% of total licensed vessels in the Straits of Malacca (Yahaya, 1988). To mitigate the economic fallout, the fishery authority decided to permit many small trawlers to continue operating until these vessels become obsolete.

Between 1981 and 2018 period, DOFM's official figures showed that licensed industrial trawlers and purse seiners have dropped by about 50%, in particular mini trawlers suffered a stiff decline of more than 90% from the peak recorded in 1981 (see Table 1). Although these new policies have greatly reduced trawl fishing fleet particularly mini trawlers, but the Malaysian government had paid little attention to the uncontrollable growth in the traditional fishing gear sector (e.g., drift/gill nets) (Stobutzki, Silvestre, Abu Talib, et al., 2006; Stobutzki, Silvestre, Garces, 2006b). In the parallel period, DOFM's official figures indicated that the pool of licensed drift/gill fishing nets had doubled from 8,453 units to 17,962 units. Though traditional gear sector had created employment and income for the marginalized poor fishermen, but the exponential growth of artisanal fishing may hinder effort reduction policy and sustainability of inshore fisheries. The design of the zoning system also failed to take into consideration the fish assemblages in these demarcated fishing regions. Garces et al. (2006) pointed out that the zoning system was primarily based on physical distances from the shoreline, which was incompatible with observed behavior patterns of demersal species. Most demersal species assemblages, in fact, were distributed across various zoning limits, which implied that fishermen operating in different zones could have targeted the same species groups in these physically delineated fishing zones. Demersal assemblages may also be influenced by other elements, such as salinity, sea temperature, seabed types, and coral reef. Over the last four decades, the Malacca Straits' inshore fisheries stock has not been restored to a healthy level but rather continued to be depleted further, and most recently the fishery authority is forced to impose a "no fishing zone" rule within 5 NM from the shoreline. More coordinated and in-depth research that incorporated the above insights is needed to inform existing input control policy for sustainability.

4.3 Excess Capacity Reduction Measures

Ahmed's et al. (2006) research showed that overcapacity was another major driver of overfishing because fisheries sector in SEA provided the very much needed job opportunities and cheap protein nourishment for the growing population. However, this could exert excessive pressure on the depleted fish stock which in turn produced greater competition among traditional and large-scale commercial fishermen for the remaining stocks. A few recent studies pointed to an overfishing syndrome: a reversal of high-to-low commercial fish landings composition ratio from 80:20 to 20:80 over the 40 years (Wong & Yong, personal communication, 2018) increased amount of trash fish and squid landings probably due to dwindling larger predator demersal species (Sany et al., 2019), and a higher number of trawl operators struggled to meet 250 tons minimum annual landings requirement – a stipulated condition for the renewal of their licenses (Nuruddin & Isa, 2013).

Though some high-priced demersal species such as grouper displayed a slightly increasing trend between 2004 and 2012, but this could be attributed to an increase in fishing efforts, rather than a recovery of the grouper population (Miat Piah et al., 2018). Besides, the overdependence of local commercial fishing fleet on foreign fishermen (about 75–80%) might lead to excessive fishing effort because they worked harder than domestic fishermen due to attractive profit-sharing scheme.

To deal with this pressing issue, the Malaysian government had drafted the National Plan of Action 1 (NPOA 1) (2004) and NPOA 2 (2015) in accordance with FAO International Plan of Action for the Management of Fishing Capacity (IPOA-capacity) to reduce overcapacity in local fisheries. DOFM was tasked with the responsibility to reduce overcapacity through persuading less successful trawl operators to relocate to less exploited offshore regions, removing licenses of underperforming fishing vessels, and implementing license moratorium for inshore fisheries, and providing an exit plan for small trawlers (below 40 tons) (Ahmad, 2011). However, the vessel decommissioning scheme was unsuccessful due to the deficiency of funding (Nuruddin & Isa, 2013). Moreover, the lack of systematic way to reduce overcapacity allowed a surge in licensed and unlicensed traditional gears, and imprecise boat specifications created an opportunity for fishermen to enhance their fishing power as well (Williams & Staples, 2010).

In the regional level, members of the SEA countries recognized that overfishing and overcapacity seriously threaten the sustainable management and conservation of fishery resources, and consequently the Southeast Asian Fisheries Development Center (SEAFDEC) was established in 1967 to support the sustainability of fisheries and aquaculture in SEA (Harlyan, 2019). This independent and autonomous regional body had advocated better management of fishing capacity and the use of responsible fishing technologies and practices. One important consensus reached in this regional agreement was that member countries needed to move away from the 'open access' to fisheries to "limited access" through rights-based fisheries and to protect the rights and benefits of inland and coastal fisheries communities (Amornpiyakrit & Siriraksophon, 2016).

4.4 Illegal, Unreported, and Unregulated Fishing

FAO estimated that about 30% of the global catches may be linked to illegal or unreported fishing activities which threatened all conservation and sustainability measures (Ghazali et al., 2022). In Malaysia, DOFM observed that "on average, there were at least 200 Thai boats crossed and fished illegally in Malaysian EEZ waters" (Yahaya, 1988). In recent report, a local newspaper claimed that close to 1 million metric tonnes of fish (USD 1.36 billion) were harvested illegally from the Malaysian waters (mainly east coast) annually by illegal foreign fishing vessels originated from Thailand, Vietnam, and Indonesia (Majid, 2017). Illegal foreign fishermen's fishing practices were detrimental to local fisheries because they exploited local marine resources indiscriminately through advanced and sophisticated technologies. The growth of IUU fishing is further aggravated by the ever-growing demand for seafood in the SEA region because many of these coastal states rely on marine resources for livelihood and food. IUU's lucrative profits and tax-free status encouraged illegal fishers to operate on a large scale and often practiced it with impunity due to poor enforcement (Ghazali et al., 2022).

To minimize IUU activities, the Malaysian government established the Malaysian Maritime Enforcement Agency (MMEA) under the Malaysian Maritime Enforcement Agency Act 2004 to enforce fisheries regulations over Malaysia's EEZ region. MMEA involved collaborative effort of various stakeholders such as the department of fisheries (DOF), department of marine park (DOMP), marine police, navy, and the Malaysian Maritime Enforcement Agency, whose main responsibilities were to conduct joint seaborne operations, air and sea patrols, and fishing vessel inspection. MMEA was supported by 100 patrol vessels and three Boston whalers for sea patrol (Yleaña & Velasco, 2012).

However, IUU is not a unilateral problem, and there is a need for cooperation among regional countries, and thus Yleaña and Velasco (2012) encouraged the establishment of regional and subregional coordination and management of fisheries resources which include the development of regional and subregional Monitoring, Control, and Surveillance (MCS). Malaysia, Singapore, and Indonesia (MALSINDO) took the initiative to set up a trilateral cooperation mechanism to conduct collaborative sea and air patrol particularly against piracy and IUU fishing in the territorial waters of the Strait of Malacca (Pedrason et al., 2016). Meanwhile, members of SEA nations also created the SEAFDEC as well as the Regional Fishing Vessels Record (RFVR). This initiative intended to construct a regional record of fishing vessels, starting with smaller vessels (24 m in length) which could be expanded later when time comes (Ghazali et al., 2022). The RFVR together with refined fishing licensing systems could be useful to rein in IUU fishing activities in Southeast Asia. Pedrason et al. (2016) commented that regional countries generally did not have a single central authority entrusted with power to regulate and enforce fisheries regulation against IUU activities. For example, both Malaysia and Indonesia had 12 authorized institutions in maritime security with overlapping roles and thus hindered bilateral collaboration on combating IUU in the Straits of Malacca.

4.5 Remedies for Weaknesses in Input Control Approach

Though zoning system intended to reduce overfishing, it failed to consider fish assemblages across different zoning regions. Garces et al.'s (2006) studies indicated that there were "shallow" and "deep" demersal fish assemblages, demarcated at about 40 m depth in the Malacca Straits. Given most fish assemblages are likely straddled across various zoning boundaries, fishermen operating in different zones might have been catching the same species groups at the same time. In view of this, more research on fish assemblages, the life cycle, and migratory patterns of demersal species are needed to prevent overfishing of the same species in the future. Similarly, neighboring nations (Malaysia, Indonesia, and Thailand) should conduct joint research on pelagic species' natural habitat and migratory routes because these marine fishes are known for their long-distance migratory habits. Such regional collaboration is better than a unilateral effort to effectively deal with the conservation of fisheries in the Straits of Malacca.

In terms of property rights, there is still an unresolved dispute about fair distribution of fishing licenses between industrial and traditional fishermen in the inshore fisheries. To resolve such problem, a nonbiased and trusted independence agency should be set up to allocate fishing rights fairly between the competing groups and to provide financial incentives to encourage less successful fishermen to seek alternative employment such as aquaculture or ecotourism. To reduce overfishing in the inshore regions, the fisheries authority needs to conduct a feasibility study on the optimum level of artisanal gears, particularly drift/gill nets, to effectively reduce the overcapacity and overfishing problems in the Malacca Straits.

As for fish net noncompliance, the main reason is because valuable small fishes and crustaceans such as small grouper and prawn are usually found in the inshore regions. On the practical side, it is hard to enforce fish net regulation because fishermen are allowed to catch all types of marine species year-round with just one license. Maybe, it is time to restructure the fishery licensing system and open sea concept so that fishermen are licensed only to catch certain types of fishes in certain time of the year. The government should consider issuing fishing license according to fish species and crustaceans, determine the optimum fishing quota, and closed season for fishing to restore the depleted fish stock. These new measures need to be based on best scientific evidence and must take into consideration the impact of such new policy on the livelihood, employment, income of local fishermen in order to elicit their support for fishery conservation policy.

5. Marine Ecosystem Conservation Policy

5.1 Conservation of Natural Habitats and Ecosystem

SEA is not only famous for the abundance of many diverse coral reef fishes and marine creatures but also possesses rich fishery habitats and ecosystem (Williams & Staples, 2010). Malaysia, for example, has extensive natural ecosystem, such as mangroves, mudflats, coral reefs, and seagrass meadows, which can be found near inshore regions and river estuaries. This ecological system helps to provide nursery grounds and refugia for the sustainable growth of local fisheries.

Malaysia's mangrove forests, for instance, are one of the best-developed and most biodiverse ecosystems in the world which help to prevent coastal erosion and control of floods and sediments (Raza et al., 2013). With an estimated 640,000 ha of mangrove forests, Malaysia is the country with the fifth largest mangrove area in the world (Sany et al., 2019). About 17% (103,000 ha) of mangrove is resided in Peninsular Malaysia and mostly along coastline and estuaries of the Malacca Straits due to the region's relatively calmer wind fetch. Besides offering protection against natural elements, mangrove forests make significant contribution to coastal fisheries in terms of providing trophic and refuge support and fish nursery. Chong (2007), a veteran of Malacca Straits' estuaries, believed that mangroves sustained more than half of the annual offshore fishery landings of 1.28 million tons. Recent research showed that abundance of shrimps, fishes, and other marine organisms could be found in healthy mangroves (Faridah-Hanum et al., 2019).

Mudflats also play an important role in the ecological system in the Straits of Malacca, and they are usually connected to mangrove forests. An estimated 11,000 million tons of cockle seeds and 72,000 tons of commercial cockles are produced annually in the Straits of Malacca. Perak's Larut Matang (40,711 ha) is home to the largest protected mangrove swamp in the Straits, mangroves of Johor (25,618 ha) and Klang (22,500 ha) take second and third place, respectively. Matang mangrove reserve provides sanctuaries for many land creatures and birds in addition to sustain more than 17 kinds of shrimp and 117 fish species (Sany et al., 2019).

Coral reefs also feature as an important element in supporting the ecological system, and they are mainly found in the Payar group of islands in Kedah and the Sembilan group of islands in Perak (Gopinath & Puvanesuri, 2006). Coral reefs shelter unprotected shorelines from wave action and tropical storms and provide both trophic support and refuge for diverse marine fishes, creatures, and organisms. However, there are less seagrass meadows in the Straits of Malacca, but seagrass provides important nursery and breeding grounds for fisheries species and marine mammals. Hashim's et al. (2017) survey showed that there were less sighting of dugong in the western areas, compared to eastern and southern areas, of the Johor Straits due to the degradation of seagrass habitat, which dugong used as feeding ground.

5.2 Pollution and Environmental Policy

The Straits of Malacca is most susceptible to ship-discharged marine pollution like oil and grease because it is a popular international shipping route for many international commercial vessels. Oil is toxic to the marine environment particularly the polycyclic aromatic hydrocarbons (PAHs), a main component of petroleum, which is hard to clean up and could reside in the sediment and marine environment for a long time. Marine creatures that are persistently exposed to PAHs can experience health and developmental problem. Kasmin (2010) reported that the number of vessels passing through the Strait had increased from 55,957 to 62,621 ships during the 2000–2005 period, and there were 108 cases of illegal discharge of oil waste out of 144 cases of oil spills into the sea; however, only 14 ships were convicted of pollution. One of the main reasons was attributed to MMEA's poor training and lack of appropriate tools to handle illegal oil discharge in open sea (Kasmin, 2010).

Similarly, PAHs pollution also occurred in certain coastal areas particularly river mouth and estuaries of Malacca Straits: low to moderate-high level of PAHs risk exposure in both Klang Strait (Tavakoly et al., 2014), Malacca and Prai rivers (Raza et al., 2013). Keshavarzifard's et al. (2017) research showed the breakdown of sources of PAHs pollution in sediment: diesel emissions (30.38%), followed by oil and oil derivatives and incomplete coal combustion (23.06%), vehicular emissions (16.43%), wood combustion (15.93%), and natural gas combustion (14.2%). The study found that the presence of PAHs in short-neck clam would increase potential carcinogenic effects in the consumers in the long term. As for other heavy metals (e.g., lead, arsenic, zinc) pollution, most recent studies provided no conclusive results: heavy metal elements in sediments were within acceptable level but relatively more serious pollution in Perak river and surrounding coastal areas (Zhang et al., 2022); no indication of serious pollution except relatively high concentration in the river mouth of Kuala Perlis (Shaari et al., 2018); acceptable levels of heavy metals presence in sediments along Selangor's Langat estuaries (Mokhtar et al., 2015); and unsafe levels of arsenic and copper in most sampling stations along the Malacca's coastline (Looi et al., 2013). In Malaysia, some irresponsible industrial manufacturers, especially illegal factories operators, paid little regard to environmental safety due to lax enforcement. A local daily recently reported that the irresponsible discharge of chemical effluents in Sungai Kim Kim caused more than 4,000 residents to fall sick, and those who were affected displayed a variety of symptoms from nausea and dizziness to vomiting (Perimbanayagam, 2019).

Several studies have also been conducted on possible seafood contamination. A study by Alina et al. (2012) examined the possible presence of toxic substance such as cadmium, lead, arsenic, and mercury in 12 local favorite marine fish species (e.g., ray, threadfin bream, grouper, Indian and Spanish mackerel) and found that the levels of heavy metals in fish species were below the permitted limit set by FAO or World Health Organization (WHO). Similarly, other recent studies on fish contamination in the Straits of Malacca indicated no unacceptable levels of harmful toxic elements in seafood, except for large demersal fish (over 20 cm in length) which contained a relatively high level of mercury but within the acceptable limits (Ahmad et al., 2015; Anual et al., 2018; Kamaruzzaman et al., 2011; Looi et al., 2016; Nor Hasyimah et al., 2011).

The fragile coastal ecosystems and natural habitats are still under serious threat by pollution caused by ongoing human activities such as the conversion of mangrove forests into aquaculture farming and tourism. For example, cockle breeding, a multimillion-ringgit industry in Malaysia's coastal areas, was almost wiped out due to sharp decrease in production, with a drastic drop from peak production of 100,000 tons in 2005 to merely 16,000 tons in 2015, which was mainly attributed to pollution (Spykerman, 2016). However, any attempts to protect and conserve ecosystems are often derailed by complex socioeconomic and political factors, which occur under the context of a multiethnic and loose federal-state relationship (Saad et al., 2013). The major drawback in environmental protection is attributed to unclear jurisdictions and the absence of a central institution empowered to protect the vulnerable and rich biodiverse ecosystems. Saad et al. (2013) opined that the DOFM should be given the lead role to revamp and to build an integrated regime to tackle the eco-environment and fisheries issues. Williams and Staples (2010) further commented that current environmental reports were mostly derived from aggregate, fragmented, and ad hoc research, instead of regular and systematic monitoring. Therefore, they proposed a new multilateral mechanism entitled, "Partnerships in Environmental

Management for the Seas of East Asia" (PEMSEA) with the aim to create a more systematic, holistic approach to coastal reporting management system. To protect the eco-system and to minimize oil discharges and ship collusions in the Straits, Malaysia, Singapore, and Indonesia have recently ratified a new multilateral mandatory ship reporting system to mitigate the pollution problem.

5.3 Marine Protected Areas and Ecotourism

Marine Protected Areas (MPAs) were an important tool that could be used to protect and sustain marine habitat and biodiversity of the ecosystems such as coral reefs (Islam et al., 2017). Malaysia's coral reef areas are estimated at about 3,600 km^2 which are subjected to the onslaughts of uncontrollable natural elements and human activities. Although Malaysia had constructed artificial reefs to rehabilitate the coral reef ecosystem in the past, some research showed that they could not generate new biomass nor totally replace the natural coral reef ecosystems (Masud & Kari, 2015; Ramli et al., 2002). About 42 MPAs had been established by DOFM in the 1980s (Islam et al., 2014), and by 1994 most coral reef islands had been proclaimed as MPAs under the Marine Parks and Marine Reserve Order 1994. In Malacca Straits, some of the largest coral reef MPAs can be found in Pulau Paya, Pangkor island, and Pulau Besar, which were established and continue to be managed by the fishery authority (Ali et al., 2013).

To better protect the status of MPAs, the Marine Park Department Malaysia (MPDM) was established by the Malaysian government in 2004 under the Ministry of Natural Resources and Environment (MNRE) (Islam et al., 2017). MPDM's primary objective is to improve the management of MPAs through effective enforcement of MPA regulations. However, the establishment of a new MPA is not as straight forward as just declaring a specific place as protected area. It often involved conflict of self-interests and strong debates among the stakeholders, particularly the local community; therefore, before designating a new establishment, community perceptions and livelihood impact study needed to be investigated probably to ensure the success of such plan (Bennett & Dearden, 2014). Lastly, MPAs could possibly help to sustain local fisheries and promote national ecotourism and financially contribute to the sustainable management of fisheries (Vianna et al., 2018).

5.4 Law Enforcement

In the early days, DOFM is entrusted to enforce fisheries regulations (e.g., Fisheries Act of 1985 and EEZ Act 1984) in the areas of input controls, ecosystems, MPAs, destructive fishing methods, and IUU fishing. However, budget and manpower constraint coupled with vast coastline hampered the ability of the agency to carry out effective supervision. Illegal foreign fishing vessels freely encroached into the EEZ zone and damaged local fishing gears and reduced the landings and income of domestic fishermen (Wong & Yong, personal communication, 2018). To solve the problem, several researchers had put forward several solutions: imposing hefty fines (Mohamed, 1991); handing out harsh fines to the hard-core habitual offenders (Ali & Hussin, 2010); and setting up a regional fishing vessel record (Ponsri et al., 2014). To improve the effectiveness of law enforcement, the Malaysian government had decided to merge diverse law enforcement agencies into a single entity in 2004, which was called the MMEA, and initiated a MCS program that was supported by a vessel monitoring system (VMS), patrol boats, and aircraft to reduce IUU and other maritime security issues in Malacca Straits (Zahaitun & Saharuddin, 2008). Around that time, MPDM was formed out of DOFM to effectively deal with IUU fishing and to protect ecosystems around MPAs regions.

Beside IUU problem, the Malaysian government must deal with sea-based pollution. One of the major sources of pollution came from oil spills caused by the offshore oil and gas companies, and to resolve such problem, these firms are required to install oil-spill response plans. In the regional level, Malaysia, Singapore, and Indonesia have also set up a mandatory multinations ship reporting system to prevent and cut down pollution.

5.5 Recommendations to Improve the Marine Ecosystem Conservation Plan

Due to anthropologic development activities over time, the size of mangrove forests has been gradually reduced in size and threatened by pollution. To deal with such issue, the Malaysian government needs to employ more comprehensive mangrove health assessment tool such as Faridah-Hanum et al.'s (2019) integrated ecological socioeconomic method. This innovative model incorporated 43 indicators from mangrove biotic integrity, mangrove soil, marine-mangrove, mangrove hydrology, and mangrove socioeconomic factors, which are relatively less laborious and cost-effective in monitoring the health of mangroves compared to traditional labor-intensive way of checking the number of timber stock. Replanting of selected mangrove species strategically along the extended coastline could strengthen the ecosystems and restore fisheries habitats along the Straits of Malacca as well. Instead of employing a top-down approach, fisheries authority needs to solicit the help of local fishermen and fisheries experts to identify suitable areas for replanting mangroves to ensure sustainable development of mangrove forests. The government can explore other useful scientific method as well such as benthic Foraminifera Stress Index (FSI), which examines how dissolved oxygen, heavy metal concentrations, and water depth could affect benthic foraminiferal distributions in the shallow tropical marine environments along the Straits of Malacca (Minhat et al., 2020).

Long standing unclear and ambiguous jurisdictions on territorial waters, especially between the state and federal governments, can hamper the enforcement of fisheries regulations along the Straits of Malacca. To resolve such problem, there is a need to set up an independent commission that composes of legal experts from both state and federal departments to fine-tune the laws to pave the way of effective management and enforcement of laws in the Malacca Straits. Given MMEA is overburdened with too many roles (e.g., national defense, maritime terrorism, maritime crimes, IUU fishing, and pollution), and so there is a need to restructure the national enforcement agency so that a subunit of enforcement agency can be created to deal specifically with noncompliance and IUU activities. The new fisheries enforcement agency could divide the long and narrow Malacca Straits into northern, central, and southern regions to monitor fishing activities. More funding and resources are needed for this kind of restructuring process. In the regional level, to reduce the conflict of overfishing rights with neighbor Indonesia, both countries can refer the overlapping claims of Malacca Straits to an independent international tribunal for a fair and just decision. This will enhance safety, facilitate future regional cooperation, and increase the effectiveness of fisheries management in the Straits of Malacca.

6. Other Factors that Contribute to Overfishing

6.1 Socioeconomic Factors

Socioeconomic factors such as population growth, consumer preference for healthier organic food, and income can further exacerbate fishing pressure on depleted fisheries and ecology. Garcia and Rosenberg (2010) reported that the world population has increased from 4.5 billion to 8 billion between 1981 and 2020 period, while Malaysian population has doubled to 33 million over the same time span. Similarly, real gross domestic product per capita has tripled over that same period. Growing income usually changes people's preference to seek after more nutritious and healthy food such as marine fish resources. Malaysia's trade statistics on fisheries products validated this presupposition. Over the 2000–2015 period, DOFM reported that exports of fishes (including freshwater fish) had doubled from 144,600 metric tons (MT) to 260,000 MT, but imports increased in parallel manner, from 323,000 MT to 444,000 MT, and trade deficit had widened from USD29 million in 2009 to USD228 million in 2015 – just over a span of seven years (DOFM, 2021).

In terms of fish price sensitivity, past research results indicated that increasing income and urbanization had direct positive impact on fish and fish products consumption (Dey et al., 2008; Tan et al., 1995). Dey et al.'s (2008) study also revealed that high-value fishes had relatively higher demand elasticity over low-value species in Malaysia. The possible explanation was that high-price species such as grouper and shrimp tend to have alternative substitutes (for example, beef and poultry products) at that high price range while cheaper fish (normally consumed by low-income group) had no cheap animal protein substitutes at that low price range (Dey et al., 2008).

Other studies propagated sustainable fisheries through ecolabeling. Given the inadequacy of Malaysia's input control, enforcement and IUU policies, Reckova et al. (2013) argued that ecolabeling might be a useful tool to mitigate the overly exploited fish population. In another study, Brecard et al. (2009) showed that young and educated consumers in Europe tend to have higher preference for ecolabelling fish products; however, Jacquet and Pauly's (2007) findings showed ecolabeled seafood products received some successes in certain regions in the

Western countries but poor response from Asian countries. Given limited studies on consumer behavior in Asian countries like Malaysia, more research on consumer behaviors is needed especially the linkage between fish demand and sustainable production. New insights are needed to shed light on how to change consumer unsustainable consumption pattern and lifestyle into a more responsible living to maintain and restore the health of marine resources in the longer run.

6.2 Government Subsidies

Inappropriate subsidy could also increase fishing efforts and lead to ever-declining fisheries stock around the world. Milazzo (1998) estimated that effort- and capacity-enhancing subsidies were in the range of USD 14–20.5 billion annually, about 20–25% of global fishing industry revenue. China, the biggest fishing nation in the world, spent USD 6.5 billion on fisheries subsidies in 2013, and 94% went to fuel subsidy. The central government also intended to further expand and modernize its distance water fishing industry through state subsidy and increase its fleet from current 1899 vessels to 2,300 ships in 2015 (Sakai et al., 2019). The goal of the fleet expansion was to reach a target offshore landing of 1.7 million tons at an estimated revenue of USD 2.9 billion. Other researchers such as Lee and Viswanathan (2019) argued that there were three types of subsidies: beneficial subsidies, ambiguous subsidies, and harmful subsidies. Beneficial subsidies helped the traditional fishermen to increase landings and incomes, to minimize bycatch, and to reduce overcapacity and excessive fishing effort – spending on fisheries management as well as research and development are a good example. In contrast, harmful subsidies encourage capacity-expansion in the fisheries that lead to overfishing, for instance, subsidy for boat construction, fleet modernization, and fuel support.

In Malaysia, state subsidies for fisheries stood at USD 120 million in the year 2017; fuel subsidies made up about 60% of total fisheries subsidies while living subsidies accounted for 31.7% of total fisheries subsidies. Ali et al. (2017) argued that fuel subsidies were bad subsidies because these subsidies were given to fishers to reduce fishing cost, but living subsidies helped to provide income support to fishers and reduce income fluctuation during monsoon season or rough water condition The same study showed the breakdown of the percentage of income earned from subsidies for different fishing zones: Zone A (38%), Zone B (66%), and Zone C (28%). Similarly, the contribution of fuel subsidy to fishermen income in different fishing zones were estimated: zone A (18%), zone B (54%), zone C (24%), which implied that fishermen were very much dependent on fuel subsidy especially zone B. Some new studies revealed new aspects of fisheries subsidies; Cullis-Suzuki and Pauly (2010) argued that MPAs, which help to conserve marine fisheries and ecology in both short and long run, can be treated as a form of beneficial subsidy for fishermen. They estimated global maintenance cost for MPAs at just USD$ 870 million which is much lower than the global subsidy of USD$ 35 billion. Other researchers such as Lee and Choi (2017) suggested that the subsidy for exported fisheries products should be abolished but financial support should be given to fisheries products that were destined for domestic consumption or processing.

Nevertheless, the reduction of harmful subsidy needs international cooperation because most fisheries straddle across different national EEZ fishing boundaries particularly migratory fisheries species. To reduce overcapacity and overfishing problem, international effort is needed to formulate regulation to reduce inappropriate fisheries subsidies (Chou & Ou, 2016). In 2007, the World Trade Organization (WTO), after lengthy negotiations, presented a draft text on fisheries subsidies; however, no final consensus came out of the WTO meeting. More research and discussion need to be carried out on harmful subsidy in the international arena to reduce overfishing not only in EEZ regions but also in the deep seas to prevent sudden stock crash worldwide.

6.3 Recommendations for Socioeconomic Problems

Today, about one out of six full-time fishermen are working in Malaysia's aquaculture sector, and they contributed about 20% of total domestic seafood production, with an economic value estimated at USD 700 million (Azra et al., 2021; FAO, 2020). However, the aquaculture farming sector still failed to achieve the national target of 50% of total seafood production by 2020. A host of factors such as inadequate government funding, disease outbreak, pollution, and lack of interested domestic workers especially youth contributed to less ideal production level. The Malaysian government needs to spend more on research funding related to disease prevention and carry out more effective and stricter enforcement against offenders to minimize toxic and heavy metal pollution in estuaries and coastal regions. Without purposeful and active government support, aquaculture is unlikely to reach its potential output and to help to reduce excessive fishing pressure in the local marine fisheries particularly inshore regions.

In addition, existing fishery management literature predominantly focuses on either supply-side fishery management or ecosystems, but there is very little research on the demand-side management in the fisheries sector. Socioeconomic factors such as growing population, increasing household income, changing in preference for healthy organic seafood, wasteful lifestyle, and other elements can contribute to unsustainable demand which will indirectly contribute to unsustainable exploitation of marine resources in the Straits of Malacca. More comprehensive and in-depth research is needed to analyze important demand factors and to produce new insights to inform sustainable demand in the future, for the sake of sustainability in commercial fisheries.

7. Complimentary Strategies

7.1 Offshore and Deep-Sea Fishing

To reduce overfishing particularly inshore fisheries, the Malaysian government has encouraged fishermen to fish in offshore regions since the mid-1980s. Historically, offshore fisheries only contributed about 20% of total marine production in the Straits of Malacca. Ahmad et al.'s (2003) trawl survey results revealed that even the Straits' offshore fisheries had been overfished in 1997. DOFM's (2021) statistics indicated that the offshore fisheries production had decreased from 138,360 tons in 2010 to just 78,367 tons in 2015.

To mitigate such problem, the Malaysian government has invited local fishermen to participate in a public–private partnership venture to explore opportunity in underexploited deep-seas fisheries regions. Initially, deep-sea expedition was a joint venture between Taiwanese tuna-fishing operator and Fisheries Development Board (or Lembaga Kemajuan Perikanan Malaysia (LKIM)) which focused on tuna fisheries in the Indian ocean particularly the Bay of Bengal. In 2001, Malaysia officially became a member of the Indian Ocean Tuna Commission (IOTC) and proceeded to construct the Malaysian international tuna port in Batu Maung, Penang, and offered generous economic incentive packages to attract Malay fishermen to join this deep-sea fishing industry (Ismail et al., 2012). Latest studies (Basir & Bakar, 2011; Kaur, 2014) indicated that Malaysian deep-sea fleets were consisted of five fishing vessels and one carrier targeting oceanic tuna (mainly *Albacore* species). About 1,300 tons, on average, were caught in the 2010–2015 period, usually uploaded, and sold off at Port Louis, Mauritius. Despite strong government financial support, the deep-sea fish landings are still far below potential target and still faced many challenges today, such as badly managed funding and facilities, and low interest among local fishermen (Ahmad, 2011).

7.2 Aquaculture

Besides deep-sea fishing, the Malaysian government has encouraged fishermen to diversify into aquaculture farming. Given the Malacca Straits' favorable climatic setting and ample sheltered marine waters, the aquaculture sector (especially brackish water aquaculture) can generate substantial benefits for local fishermen, such as employment, income, and fish production. Malaysia has earmarked 159,633 ha of potential aquaculture land bank, but only 10% of it has been employed for aquaculture farming (Ahmad et al., 2003). In 2015, DOFM (2015) reported that a total of 23,832 marine aquafarmers were employed to produce 100,000 tons of marine fishes and prawn, mainly grouper, seabass, and tiger prawn. In addition, inland freshwater aquaculture, primarily pond culture, produced about 105,000 tons of fish at an estimated value of USD 168 million. The main cultured species of freshwater fish include Nile tilapias (44.7%), catfish (36.7%), carps (10.08%), and other species (8.52%) (Kurniawan et al., 2021). High-value species of freshwater aquaculture are composed of marble goby, masheer fish empurau, red tilapia, and silver catfish in addition to seaweed production of about 174,083 tons valued at USD 12 million.

Although presently aquaculture production has reached about 20% of total fish production in Malaysia, it is still far below the national target of a national output at 50% (Wong & Yong, 2020). Besides underperformance, there are several issues faced by the aquafarmers such as effluents, early mortality syndrome diseases, overreliance on fishmeal, inadequate financial support from government, and pollution. For example, a local daily reported that severe heavy metal pollution, particularly nickel, occurred at Penang's Teluk Bahang and Nibong Tebal's river mouth – about 1,000 times more than the normal measurement of 0.005 ppm in typical sea water (Sekaran, 2019). As a result, fish farmers suffered multimillions of dollar losses due to the sea-water pollution. Massive wastewaters, mainly produced by man-made ponds in brackish water and freshwater cultures, are also becoming another concern. Aquaculture's effluent is harmful to the environment because it carries many pollutants such as high organic concentrations, nutrients, proteins, hormones, as well as biomasses (Dauda et al., 2019). Kurniawan et al. (2021) proposed the use of wastewater technology to recover nutrient-rich solids and solutions through the coagulation–flocculation process utilizing natural, or bioflocculants accompanied by constructed wetlands using floating plants. The residues generated from the coagulation–flocculation process can be transformed into value-added products, like soil conditioners or fertilizers, while soluble nutrients in the floating plants can be recycled as feedstocks for fish breeding. Lastly, the treated effluent free from pollutants can be reused for fish farming. This form of green technology can help to promote sustainable environments.

8. Conclusions

Historically, the Straits of Malacca is blessed with plentiful fishes and crustaceans; however, after the introduction of commercial trawlers and purse seines, fishing practice has changed from subsistence to commercial fishing. In just over a short span of time, marine fishes, especially coastal fisheries, have been overfished, which prompted the Malaysian government to implement a legal framework to preserve and restore the overexploited fisheries. The comprehensive and systematic analysis of fishery management literature on input controls, ecosystems, enforcement, aquaculture, and socioeconomic factors has indicated that there are rooms for improvement for existing conservation framework. Greater attention need to be given to some highlighted issues such as the need to resolve unclear and conflicting national and international territorial waters rights, strengthen national and multilateral research collaboration on fish life cycle and migratory routes, employ effective and cost-saving ecological tools to identify and expand MPAs areas, restructure MMEA to deal more effectively with noncompliance and IUU issues, and promote sustainable demand for marine fisheries particularly endangered fish species. With more systematic and focus research, there is hope that more insightful ideas and creative solutions can be brought forward to establish a more effective and sustainable marine fisheries management system in the future.

References

Abdullah, N. M. R. (1995). Towards an integrated fisheries information system in Malaysia. *Marine Resource Economics, 10*(3), 312–320. https://doi.org/10.1086/mre.10.3.42629594

Ahmad, M. Z. (2011). *International legal and normative framework for responsible fisheries, with reference to Malaysia's offshore EEZ fisheries management.* (Doctoral dissertation). https://ro.uow.edu.au/cgi/viewcontent.cgi?referer=https://www.google.com/&httpsredir=1&article=4410&context=theses

Ahmad, A. T., Isa, M. M., Ismail, M. S., & Yusof, S. (2003). Status of demersal fishery resources of Malaysia. In G. Silvestre, L. Garces, I. Stobutzki, M. Ahmed, R. A. Valmonte-Santos, C. Luna, L. Lachica-Aliño, P. Munro, V. Christensen, & D. Pauly (Eds.), *Assessment, management and future directions for coastal fisheries in Asian countries.* WorldFish center conference proceedings (pp. 83–136). World Fish Center.

Ahmad, N. I., Noh, M. F. M., Mahiyuddin, W. R. W., Jaafar, H., Ishak, I., Azmi, W. N. F. W., Veloo, Y., & Hairi, M. H. (2015). Mercury levels of marine fish commonly consumed in Peninsular Malaysia. *Environmental Science and Pollution Research, 22*(5), 3672–3686. https://doi.org/10.1007/s11356-014-3538-8

Ahmed, M., Salayo, N. D., Viswanathan, K. K., Garces, L. R., & Pido, M. D. (2006). *Management of fishing capacity and resource use conflicts in Southeast Asia: A policy brief.* World fish center.

Alam, M. F., Omar, I. H., & Squires, D. (2002). Sustainable fisheries development in the tropics: Trawlers and licence limitation in Malaysia. *Applied Economics, 34*(3), 325–337. https://doi.org/10.1080/00036840110036305

Ali, J., Abdullah, H., Noor, M. S. Z., Viswanathan, K. K., & Islam, G. N. (2017). The contribution of subsidies on the welfare of fishing communities in Malaysia. *International Journal of Economics and Financial Issues, 7*(2), 641–648.

Ali, J., Ariff, S., Viswanathan, K. K., & Islam, R. (2013). Effectiveness of marine protected areas as a management tool for the management of the seas of Malaysia. *Australian Journal of Basic and Applied Sciences, 7*(8), 658–666. https://d1wqtxts1xzle7.cloudfront.net/47325051/Effectiveness_of_Marine_Protected_Areas_20160718-16335-j4ak3m.pdf

Ali, J., & Hussin, A. (2010). Impact of enforcement and co-management on compliance behavior of fishermen. *International Journal of Economics and Finance, 2*(4), 113–121. https://www.researchgate.net/profile/Jamal_Ali/publication/47456961_Impact_of_Enforcement_and_Co-Management_on_Compliance_Behavior_of_Fishermen/links/02bfe50fe814857f26000000.pdf

Alina, M., Azrina, A., Mohd Yunus, A., Mohd Zakiuddin, S., Mohd Izuan Effendi, H., & Muhammad Rizal, R. (2012). Heavy metals (mercury, arsenic, cadmium, plumbum) in selected marine fish and shellfish along the Straits of Malacca. *International Food Research Journal, 19*(1), 135–140.

Amornpiyakrit, T., & Siriraksophon, S. (2016). Management of fishing capacity for sustainable fisheries: RPOA-capacity. *Fish for the People, 14*(2), 18–23.

Anual, Z. F., Maher, W., Krikowa, F., Hakim, L., Ahmad, N. I., & Foster, S. (2018). Mercury and risk assessment from consumption of crustaceans, cephalopods, and fish from West Peninsular Malaysia. *Microchemical Journal, 140*, 214–221. https://doi.org/10.1016/j.microc.2018.04.024

Azra, M. N., Kasan, N. A., Othman, R., Noor, G. A. G. R., Mazelan, S., Jamari, Z., Gianluca, S., & Ikhwanuddin, M. (2021). Impact of COVID-19 on aquaculture sector in Malaysia: Findings from the first national survey. *Aquaculture Reports, 19*, 100568. https://doi.org/10.1016/j.aqrep.2020.100568

Basir, S., & Bakar, N. A. (2011). Analysis of catch of neritic tuna and sharks in Malacca Strait, west coast of Malaysia peninsula. https://www.iotc.org/documents/analysis-catch-neritic-tuna-and-sharks-malacca-strait-westcoast-malaysia-peninsula

Bennett, N. J., & Dearden, P. (2014). From measuring outcomes to providing inputs: Governance, management, and local development for more effective marine protected areas. *Marine Policy*, *50*, 96–110. https://doi.org/10.1016/j.marpol.2014.05.005

Blaber, S. J. M., Cyrus, D. P., Albaret, J. J., Chong, V. C., Day, J. W., Elliott, M., Fonseca, M. S., Hoss, D. E., Orensanz, J., Potter, I. C., & Silvert, W. (2000). Effects of fishing on the structure and functioning of estuarine and nearshore ecosystems. *ICES Journal of Marine Science/Journal du Conseil*, *57*(3), 590–602. https://doi.org/10.1006/jmsc.2000.0723

Brécard, D., Hlaimi, B., Lucas, S., Perraudeau, Y., & Salladarré, F. (2009). Determinants of demand for green products: An application to eco-label demand for fish in Europe. *Ecological Economics*, *69*(1), 115–125. https://doi.org/10.1016/j.ecolecon.2009.07.017

Broadhurst, M. K., Brand, C. P., & Kennelly, S. J. (2012). Evolving and devolving bycatch reduction devices in an Australian penaeid-trawl fishery. *Fisheries Research*, *113*(1), 68–75. https://doi.org/10.1016/j.fishres.2011.09.006

Butcher, J. (1996). The marine fisheries of the western archipelago towards an economic history 1850–1960's. In D. Pauly & P. Martosubroto (Eds.), *Baseline studies of biodiversity: The fish resources of western Indonesia* (pp. 24–39). International Center for Living Aquatic Resources Management.

Chong, V. C. (2007). Mangroves-fisheries linkages—The Malaysian perspective. *80*(3), 755–772. https://www.ingentaconnect.com/content/umrsmas/bullmar/2007/00000080/00000003/art00019

Chou, Y., & Ou, C.-H. (2016). The opportunity to regulate domestic fishery subsidies through international agreements. *Marine Policy*, *63*, 118–125. https://doi.org/10.1016/j.marpol.2015.09.027

Cullis-Suzuki, S., & Pauly, D. (2010). Marine protected area costs as "beneficial" fisheries subsidies: A global evaluation. *Coastal Management*, *38*(2), 113–121.

Dauda, A. B., Ajadi, A., Tola-Fabunmi, A. S., & Akinwole, A. O. (2019). Waste production in aquaculture: Sources, components, and managements in different culture systems. *Aquaculture and Fisheries*, *4*(3), 81–88. https://doi.org/10.1016/j.aaf.2018.10.002

Department of Fisheries Malaysia. (2015). National plan of action for the management of fishing capacity in Malaysia (plan 2). https://www.dof.gov.my/dof2/resources/user_1/UploadFile/Penerbitan/Senarai%20Penerbitan/NPOA.pdf

Department of Fisheries Malaysia (DOFM). (2021). Annual fisheries statistics 1970–2020. https://www.dof.gov.my

Dey, M. M., Garcia, Y. T., Kumar, P., Piumsombun, S., Haque, M. S., Li, L. P., Radam, A., Senaratne, A., Khiem, N. T., & Koeshendrajana, S. (2008). Demand for fish in Asia: A cross-country analysis. *The Australian Journal of Agricultural and Resource Economics*, *52*(3), 321–338.

Faridah-Hanum, I., Yusoff, F. M., Fitrianto, A., Ainuddin, N. A., Gandaseca, S., Zaiton, S., Kamarudina, N., Nurhidayua, S., Kassima, M. R., Hakeem, K. R., Ibrahim, S., Adnan, I., Abdul-Ghania, N. A., Parmana, R. P., Razakd, S. B. A.,

Ibrahima, S. A., Hilaluddind, F., Ramlia, F., & Nik-Zaidin, N. H. (2019). Development of a comprehensive mangrove quality index (MQI) in Matang mangrove: Assessing mangrove ecosystem health. *Ecological Indicators*, *102*, 103–117. https://doi.org/10.1016/j.ecolind.2019.02.030

Food and Agriculture Organization. (2020). Fishery and aquaculture country profiles. *Malaysia*. http://www.fao.org/fishery/facp/MYS/en

Garces, L. R., Stobutzki, I., Alias, M., Campos, W., Koongchai, N., Lachica-Alino, L., Mustafa, G., Nurhakim, S., Srinath, M., & Silvestre, G. (2006). Spatial structure of demersal fish assemblages in South and Southeast Asia and implications for fisheries management. *Fisheries Research*, *78*(2–3), 143–157. https://doi.org/10.1016/j.fishres.2006.02.005

Garcia, S. M., & Rosenberg, A. A. (2010). Food security and marine capture fisheries: Characteristics, trends, drivers and future perspectives. *Philosophical Transactions of the Royal Society B: Biological Sciences*, *365*(1554), 2869–2880.

Ghazali, F., Talaat, W. I., Rahman, A., & Rusli, H. (2022). Malaysian efforts in combating IUU fishing: A legal and policy review. *Journal of East Asia and International Law*, *12*(2), 387–400.

Gopinath, N., & Puvanesuri, S. S. (2006). Marine capture fisheries. *Aquatic Ecosystem Health and Management*, *9*(2), 215–226. https://doi.org/10.1080/14634980 600721086

Graham, M. (1935). Modern theory of exploiting a fishery, and application to north sea trawling. *ICES Journal of Marine Science*, *10*(3), 264–274. https://doi.org/10.1093/icesjms/10.3.264

Haddon, M. (2011). *Modelling and quantitative methods in fisheries* (2nd ed.). Chapman & Hall.

Harlyan, L. I. (2019). *Purse seine fishery management in Malaysia: An output control for sustainable fisheries*. (Doctoral dissertation). https://eprints.lib.hokudai.ac.jp/dspace/bitstream/2115/80364/1/Ledhyane_Ika_Harlyan.pdf

Hashim, M., Ito, S., Numata, S., Hosaka, T., Hossain, M. S., Misbari, S., Yahya, N. N., & Ahmad, S. (2017). Using fisher knowledge, mapping population, habitat suitability and risk for the conservation of dugongs in Johor Straits of Malaysia. *Marine Policy*, *78*, 18–25. http://eprints.utm.my/id/eprint/76911/

Hilborn, R., & Walters, C. (1992). *Quantitative fisheries stock assessment: Choice, dynamics and uncertainty*. Chapman & Hall.

Islam, G. M. N., Noh, K. M., Sidique, S. F., Noh, A. F. M., & Ali, A. (2014). Economic impacts of artificial reefs on small-scale fishers in peninsular Malaysia. *Human Ecology*, *42*(6), 989–998.

Islam, G. M. N., Noh, K. M., & Yew, T. S. (2011). Measuring productivity in fishery sector of peninsular Malaysia. *Fisheries Research*, *108*(1), 52–57. https://doi.org/10.1016/j.fishres.2010.11.020

Islam, G. M. N., Tai, S. Y., Kusairi, M. N., Ahmad, S., Aswani, F. M. N., Senan, M. K. A. M., & Ahmad, A. (2017). Community perspectives of governance for effective management of marine protected areas in Malaysia. *Ocean & Coastal Management*, *135*, 34–42. https://doi.org/10.1016/j.ocecoaman.2016.11.001

Ismail, N., Ahmad, M., Hussain, M., & Abdul Malik, A. (2012). Large offshore remote system for Malaysian deep-sea fishing. https://www.academia.edu/3406264/Large_Offshore_Remote_System_for_Malaysian_Deep_Sea_Fishing

Jacquet, J. L., & Pauly, D. (2007). The rise of seafood awareness campaigns in an era of collapsing fisheries. *Marine Policy, 31*(3), 308–313. https://doi.org/10.1016/j. marpol.2006.09.003

Jagerroos, S. (2016). Assessment of living resources in the straits of Malacca, Malaysia: Case study. *Journal of Aquaculture & Marine Biology, 4*(1), 00070. https://doi.org/10.15406/jamb.2016.04.00070

Kamaruzzaman, B. Y., Rina, Z., John, B. A., & Jalal, K. C. A. (2011). Heavy metal accumulation in commercially important fishes of southwest Malaysian coast. *Research Journal of Environmental Sciences, 5*(6), 595–602. https://doi.org/10.3923/rjes.2011

Kasmin, S. (2010). Enforcing ship-based marine pollution for cleaner sea in the Strait of Malacca. *Environment Asia, 3*, 61–65.

Kaur, C. R. (2014). Sustainable development of marine living and non-living resources. In M. A. A. Mohamad (Ed.), *The paradox of the straits of Malacca: Balancing priorities for a sustainable waterway* (pp. 159–207). Maritime Institute of Malaysia. https://www.researchgate.net/profile/Cheryl_Kaur/publication/311924890_Sustainabl e_development_of_marine_living_and_nonliving_resources/links/59257058458515e3d43da575/Sustainable-development-ofmarine-living-and-non-living-resources.pdf

Keshavarzifard, M., Zakaria, M. P., & Sharifi, R. (2017). Ecotoxicological and health risk assessment of polycyclic aromatic hydrocarbons (PAHs) in short-neck clam (Paphia undulata) and contaminated sediments in Malacca Strait, Malaysia. *Archives of Environmental Contamination and Toxicology, 73*(3), 474–487.

Kirkley, J. E., Squires, D., Alam, M. F., & Ishak, H. O. (2003). Excess capacity and asymmetric information in developing country fisheries: The Malaysian purse seine fishery. *American Journal of Agricultural Economics, 85*(3), 647–662. https://doi.org/10.1111/1467-8276.00462

Kurniawan, S. B., Ahmad, A., Rahim, N. F. M., Said, N. S. M., Alnawajha, M. M., Imron, M. F., Abdullah, S. R. S., Othman, A. R., Ismail, N. I., & Hasan, H. A. (2021). Aquaculture in Malaysia: Water-related environmental challenges and opportunities for cleaner production. *Environmental Technology & Innovation, 24*, 101913.

Lee, C., & Choi, S. D. (2017). On the direction of fisheries subsidies programs in Korea under fortifying international regulations for fisheries subsidies. *Journal of the Korean Society of Fisheries and Ocean Technology, 53*(4), 456–470.

Lee, W. C., & Viswanathan, K. K. (2019). Subsidies in the fisheries sector of Malaysia: Impact on resource sustainability. *Review of Politics and Public Policy in Emerging Economies, 1*(2), 79–85.

Looi, L. J., Aris, A. Z., Haris, H., Yusoff, F. M., & Hashim, Z. (2016). The levels of mercury, methylmercury and selenium and the selenium health benefit value in grey-eel catfish (Plotoscanius) and giant mudskipper (Periophthalmodon schlosseri) from the Strait of Malacca. *Chemosphere, 152*, 265–273. https://doi.org/10.1016/j.chemosphere.2016.02.126

Looi, L. J., Aris, A. Z., Johari, W. L. W., Mohamad, F. Y., & Hashim, Z. (2013). Baseline metals pollution profile of tropical estuaries and coastal waters of the Straits of Malacca. *Marine Pollution Bulletin, 74*(1), 471–476. https://doi.org/10.1016/j.marpolbul.2013.06.008

Majid, N. (2017). Malaysia loses RM6b annually due to illegal fishing in South China Sea. https://www.nst.com.my/news/nation/2017/06/250427/malaysia-loses-rm6b-annually-due-illegal-fishing-south-china-sea

Martosubroto, P. (2002). Towards strengthening coastal fisheries management in South and Southeast Asia. https://agris.fao.org/agris-search/search.do?recordID=XF2016017751

Masud, M. M., & Kari, F. B. (2015). Community attitudes towards environmental conservation behaviour: An empirical investigation within MPAs, Malaysia. *Marine Policy, 52*, 138–144. https://doi.org/10.1016/j.marpol.2014.10.015

Matsushita, Y., & Ali, R. (1997). Investigation of trawl landings for the purpose of reducing the capture of non-target species and sizes of fish. *Fisheries Research, 29*(2), 133–143. https://doi.org/10.1016/S0165-7836(96)00534-6

Miat Piah, M. R., Abdul Kadir, N. H., Kamaruddin, S. A., Azaman, M. N., & Ambak, M. A. (2018). Analysis of historical landing data to understand the status of grouper populations in Malaysia. *Malaysian Applied Biology, 47*(3), 49–58. https://www.researchgate.net/profile/Rumeaida_Mat_Piah/publication/326683588_Analysis_of_historical_landing_data_to_understand_the_status_of_grouper_populations_in_Malaysia/links/5ba0bf8aa6fdccd3cb5f34cb/Analysis-of-historicallanding-data-to-understand-the-status-of-grouper-populations-in-Malaysia.pdf

Milazzo, M. (1998). *Subsidies in world fisheries: A re-examination.* World Bank.

Minhat, F. I., Shaari, H., Razak, N. S. A., Satyanarayana, B., Saelan, W. N. W., Yusoff, N. M., & Husain, M. L. (2020). Evaluating performance of foraminifera stress index as tropical-water monitoring tool in Strait of Malacca. *Ecological Indicators, 111*, 106032. https://doi.org/10.1016/j.ecolind.2019.106032

Mohamed, M. I. H. (1991). National management of Malaysian fisheries. *Marine Policy, 15*(1), 2–14. https://doi.org/10.1016/0308-597X(91)90038-D

Mokhtar, N. F., Aris, A. Z., Praveena, S. M. (2015). Preliminary study of heavy metal (Zn, Pb,Cr, Ni) contaminations in Langat River Estuary. Selangor. *Procedia Environmental Sciences, 30*, 285–290. https://doi.org/10.1016/j.proenv.2015.10.051

Nor Hasyimah, A. K., James Noik, V., Teh, Y. Y., Lee, C. Y., & Pearline Ng, H. C. (2011). Assessment of cadmium (Cd) and lead (Pb) levels in commercial marine fish organs between wet markets and supermarkets in Klang Valley, Malaysia. *International Food Research Journal, 18*(2), 795–802. https://www.researchgate.net/profile/Victor_James_Noik/publication/263088698_Assessment_of_cadmium_Cd_and_lead_Pb_levels_in_commercial_marine_fish_organs_between_wet_markets_and_supermarkets_in_Klang_Valley_Malaysia/links/00b7d53acd81d4d08e000000/Assessment-of-cadmium-Cd-and-lead-Pb-levels-in-commercial-marine-fish-organsbetween-wet-markets-and-supermarkets-in-Klang-Valley-Malaysia.pdf

Nuruddin, A. A., & Isa, S. M. (2013). Trawl fisheries in Malaysia – Issues, challenges, and mitigating measures. *Paper presented at the APFIC Regional Expert Workshop on Tropical Trawl Fishery Management,* Phuket, Thailand, 30 September to 4 October. http://www.fao.org/3/a-bo084e.pdf

Pauly, D., & Chua, T. E. (1988). The overfishing of marine resources: Socioeconomic background in Southeast Asia. *Ambio, 17*(3), 200–206. https://hdl.handle.net/20.500.12348/3246

Pedrason, R., Kurniawan, Y., & Purwasandi, P. (2016). Handling of illegal, unreported and unregulated (IUU) fishing. *Jurnal Pertahanan, 2*(1), 71–90.

Perimbanayagam, K. (2019). Crackdown on the Pasir Gudang factories dumping chemical waste illegally. https://www.nst.com.my/news/nation/2019/06/499728/crackdown-pasir-gudang-factories-dumping-chemical-wasteillegally

Perrings, C. (2016). The economics of the marine environment: A review. *Environmental Economics and Policy Studies, 18*(3), 277–301. https://doi.org/10.1007/s10018-016-0149-2

Pongsri, C., Kawamura, H., Siriraksophon, S., & Chokesanguan, B. (2014). Regional fishing vessels record: Option to mitigate IUU fishing in Southeast Asia. http://repository.seafdec.org/bitstream/handle/20.500.12066/935/SP12-1%20rfvr.pdf?sequence=1&isAllowed=y

Ramli, N., Ahmad, A., & Karim, K. (2002). Marine parks Malaysia: Current status and prospect of marine protected areas in peninsular Malaysia. In *Proceedings of IUCN/WCPA-EA-4 Taipei Conference,* Taipei, Taiwan, March 18–23. http://www.reefbase.org/resource_center/publication/pub_22103.aspx

Raza, M., Zakaria, M. P., Hashim, N. R., Yim, U. H., Kannan, N., & Ha, S. Y. (2013). Composition and source identification of polycyclic aromatic hydrocarbons in mangrove sediments of peninsular Malaysia: Indication of anthropogenic input. *Environmental Earth Sciences, 70*(6), 2425–2436.

Reczkova, L., Sulaiman, J., & Bahari, Z. (2013). Some issues of consumer preferences for eco-labeled fish to promote sustainable marine capture fisheries in Peninsular Malaysia. *Procedia – Social and Behavioral Sciences, 91*(Suppl. C), 497–504. https://doi.org/10.1016/j.sbspro.2013.08.447

Richardson, K., Asmutis-Silvia, R., Drinkwin, J., Gilardi, K. V. K., Giskes, I., Jones, G., & Hogan, E. (2019). Building evidence around ghost gear: Global trends and analysis for sustainable solutions at scale. *Marine Pollution Bulletin, 138,* 222–229. https://doi.org/10.1016/j.marpolbul.2018.11.031

Saad, J., Hiew, K., & Gopinath, N. (2013). Review of Malaysian laws and policies in relation to the implementation of ecosystem approach to fisheries management in Malaysia. http://cticff.org/sites/default/files/resources/39_Review%20of%20Malaysian%20Law

Saharuddin, A. H. (1995). Development and management of Malaysian marine fisheries: Technical conservation measures. *Marine Policy, 19*(2), 115–126. https://doi.org/10.1016/0308-597X(94)00011-G

Sakai, Y., Yagi, N., & Sumaila, U. R. (2019). Fishery subsidies: The interaction between science and policy. *Fisheries Science, 85*(3), 439–447.

Sany, S. B. T., Tajfard, M., Rezayi, M., Rahman, M. A., & Hashim, R. (2019). The West coast of peninsular Malaysia. In C. Sheppard (Ed.), *World seas: An environmental evaluation* (pp. 437–458). Academic Press.

Sekaran, R. (2019). Serious water pollution in Penang. https://www.thestar.com.my/news/nation/2019/05/28/serious-water-pollution-inpenang/

Shaari, H., Abdul Razak, N. S., Mohd Khalik, W. M. A. W., Shazili, N. A. M., & Bidai, J. (2018). Spatial distribution of heavy metals in tropical coastal sediment of the Northern Malacca Strait, Malaysia. *Nature Environment and Pollution Technology, 17*(4), 1115–1123. https://www.researchgate.net/profile/Hasrizal_Shaari/publication/329415721_Spatial_Distribution_of_Heavy_Metals_in_Tropical_Coastal_Sediment_of_the_Northern_Malacca_Strait_Malaysia/links/5c07819e928 51c6ca1ff28ca/Spatial-Distribution-of-Heavy-Metals-in-Tropical-Coastal-Sediment-of-the-Northern-Malacca-Strait-Malaysia.pdf

Spykerman, N. (2016). Cockle trade threatened. https://www.thestar.com.my/news/nation/2016/05/17/cockle-trade-threatened-productiondown-to-16000-tonnes-from-100000-tonnes/

Stobutzki, I. C., Silvestre, G. T., Abu Talib, A., Krongprom, A., Supongpan, M., Khemakorn, P., Armada, N., & Garces, L. R. (2006a). Decline of demersal coastal fisheries resources in three developing Asian countries. *Fisheries Research, 78*(2–3), 130–142. https://doi.org/10.1016/j.fishres.2006.02.004

Stobutzki, I. C., Silvestre, G. T., & Garces, L. R. (2006b). Key issues in coastal fisheries in South and Southeast Asia, outcomes of a regional initiative. *Fisheries Research, 78*(2–3), 109–118. https://doi.org/10.1016/j.fishres.2006.02.002

Tan, A. K., Yen, S. T., & Hasan, A. R. (2015). At-home consumption of fish products in Malaysia: An analysis of household expenditure survey data. *Marine Resource Economics, 30*(4), 417–433.

Tavakoly Sany, S. B., Hashim, R., Salleh, A., Rezayi, M., Mehdinia, A., & Safari, O. (2014). Polycyclic aromatic hydrocarbons in coastal sediment of Klang Strait, Malaysia: Distribution pattern, risk assessment and sources. *PLoS One, 9*(4), e94907.

Vianna, G. M. S., Meekan, M. G., Rogers, A. A., Kragt, M. E., Alin, J. M., & Zimmerhackel, J. S. (2018). Shark-diving tourism as a financing mechanism for shark conservation strategies in Malaysia. *Marine Policy, 94*, 220–226. https://doi.org/10.1016/j.marpol.2018.05.008

Williams, M. J., & Staples, D. (2010). Southeast Asian fisheries. In R. Q. Grafton, R. Hilborn, D. Squires, M. Tait, & M. J. Williams (Eds.), *Handbook of marine fisheries conservation and management* (pp. 243–257). Oxford University Press.

Wong, H. S., & Yong, C. C. (2020). Fisheries regulation: A review of the literature on input controls, the ecosystem, and enforcement in the straits of Malacca of Malaysia. *Fisheries Research 230*,105682. https://doi.org/10.1016/j.fishres.2020.105682

WWFMy. (2013). An assessment of fisheries and marine ecosystem in peninsular Malaysia. www.saveourseafood.my/wp-content/uploads/.../an_assessment_of_fisheries.pdf

Yahaya, J. (1988). Fishery management and regulation in peninsular Malaysia: Issues and constraints. *Marine Resource Economics, 5*(2), 83–98. https://doi.org/10.1086/mre.5.2.42628923

Yahaya, J., & Abdullah, N. M. R. (1993). Fisheries resources under stress: The Malaysian experience. *Paper presented at the International Association for the Study of Common Property,* Fourth Annual Common property Resource Conference, Manila, Phillipine.

Yleaña, J. S., & Velasco, P. E. L. (2012). Monitoring, control and surveillance (MCS) in Southeast Asia: Review of the establishment of regional MCS network. *Fish for the People, 10*(1), 7–12.

Zahaitun, M. Z., & Saharuddin, A. H. (2008). The Management of deep-Sea fishing in Malaysia: Realising the FAO's guidelines. https://www.academia.edu/4212866/THE_MANAGEMENT_OF_DEEP_SEA_FISHING_IN_MALAYSIA_REALISING_THE_FAO_S_GUIDELNES

Zhang, H., Liu, S., Ai, L., Cao, P., Wu, K., Cui, J., Wang, H., Mohamed, C. A. R., & Shi, X. (2022). Distribution and assessment of heavy metal in sediments of Malacca Strait. Marine. *Pollution Bulletin, 178*, 113575. https://doi.org/10.1016/j.marpolbul.2022.113575

Index